Making Urban Theory

This book facilitates more careful engagement with the production, politics and geography of knowledge as scholars create space for the inclusion of southern cities in urban theory.

Making Urban Theory addresses debates of the past fifty years regarding whether and why scholars should conceptualize southern cities as different and argues for the continued importance of unlearning existing theory. With examples from the urban question to environmental justice, urban infrastructure to basic income, this volume highlights the limitations of existing explanations as well as how thinking from the south entails more than collecting data in new places. Throughout the book, instances of juxtapositions, unease, unlearning and learning anew emphasize how theory-making from southern cases can open avenues to more creative possibilities. The book pulls theories apart, examining distinct components to better understand the universality and provinciality of empirical phenomena, causality and norms, including questions of what a city is and ought to be.

This book delivers a clearer articulation of ongoing debates and future possibilities for southern urban scholarship, and it will thus be relevant for both scholars and students of Urban Studies, Urban Theory, Urban Geography, Research Methods in Geography, Postcolonial/Southern Cities and Global Cities at graduate and post-graduate levels.

Mary Lawhon is Assistant Professor in the Department of Geography and Environmental Sustainability, University of Oklahoma, USA. Her research interests include urban theory, postcolonial studies, urban political ecology and the politics of urban infrastructure.

With contributions from:

Lené Le Roux is a doctoral candidate in the Department of Geography and Environmental Sustainability, University of Oklahoma, USA. Trained as an urban planner, her research interests include southern urban theory, place and place-claiming, critical pedagogy and South African cities.

Anesu Makina is a doctoral candidate in the Department of Geography and Environmental Sustainability, University of Oklahoma, USA. She previously worked as an environmental researcher at the University of Pretoria, South Africa. Her research interests include African cities, urban informality and political ecology.

Yaffa Truelove is Assistant Professor in the Department of Geography, University of Colorado-Boulder, USA. Her research primarily examines the connections between urban waterscapes and socio-political processes in cities of the global south, including differing regimes and institutions of everyday urban governance.

Routledge Research on Decoloniality and New Postcolonialisms
Series Editor: Mark Jackson, Senior Lecturer in Postcolonial Geographies, School of Geographical Sciences, University of Bristol, UK.

Routledge Research on Decoloniality and New Postcolonialisms is a forum for original, critical research into the histories, legacies, and life-worlds of modern colonialism, post-colonialism, and contemporary coloniality. It analyses efforts to decolonise dominant and damaging forms of thinking and practice, and identifies, from around the world, diverse perspectives that encourage living and flourishing differently. Once the purview of a postcolonial studies informed by the cultural turn's important focus on identity, language, text and representation, today's resurgent critiques of coloniality are also increasingly informed, across the humanities and social sciences, by a host of new influences and continuing insights for different futures: indigeneity, critical race theory, relational ecologies, critical semiotics, posthumanisms, ontology, affect, feminist standpoints, creative methodologies, post-development, critical pedagogies, intercultural activisms, place-based knowledges, and much else. The series welcomes a range of contributions from socially engaged intellectuals, theoretical scholars, empirical analysts, and critical practitioners whose work attends, and commits, to newly rigorous analyses of alternative proposals for understanding life and living well on our increasingly damaged earth.

This series is aimed at upper-level undergraduates, research students and academics, appealing to scholars from a range of academic fields including human geography, sociology, politics and broader interdisciplinary fields of social sciences, arts and humanities.

Arendt, Fanon, and Political Violence in Islam
Patrycja Sasnal

Populism and Postcolonialism
Edited by Adrián Scribano, Maximiliano E. Korstanje and Freddy Timmermann

National Security and Policy in America
Immigrants, Crime, and the Securitization of the Border
Wesley S. McCann and Francis D. Boateng

Making Urban Theory
Learning and Unlearning through Southern Cities
Mary Lawhon

For more information about this series, please visit: https://www.routledge.com/Routledge-Research-in-New-Postcolonialisms/book-series/RRNP

Making Urban Theory
Learning and Unlearning through Southern Cities

Mary Lawhon

With Contributions from Lené Le Roux, Anesu Makina and Yaffa Truelove

Routledge
Taylor & Francis Group

LONDON AND NEW YORK

First published 2020
by Routledge
2 Park Square, Milton Park, Abingdon, Oxon OX14 4RN

and by Routledge
605 Third Avenue, New York, NY 10017

First issued in paperback 2021

Routledge is an imprint of the Taylor & Francis Group, an informa business

Publisher's Note
The publisher has gone to great lengths to ensure the quality of this reprint but points out that some imperfections in the original copies may be apparent.

British Library Cataloguing-in-Publication Data
A catalogue record for this book is available from the British Library

Library of Congress Cataloging-in-Publication Data
A catalog record has been requested for this book

ISBN 13: 978-1-03-223885-2 (pbk)
ISBN 13: 978-0-367-34492-4 (hbk)

Typeset in Times New Roman
by Aptara, India

For JP

Contents

Tables

Acknowledgements

In the years between when I first imagined writing this book and now, the idea of being able to acknowledge those who have contributed to the journey sparked a renewed commitment to getting this work into print.

First, a specific thanks to my collaborators in various chapters: Yaffa Truelove, Anesu Makina and Lené Le Roux. While much of this book is focused on my own journey, it is born from many heads, hands and hearts. I am indebted to you all for the insights and pushes at critical times. I would not have gotten to this point alone.

My time in South Africa, and my commitment to (re)thinking through and from the global south, has been deeply shaped by Muna Lakhani. We met over coffee when I was twenty-one, a few days into my first sojourn outside the United States. I subsequently spent six months as a volunteer for the environmental justice organization Earthlife Africa in the leadup to the 2002 World Summit on Sustainable Development in Johannesburg. Muna's tutorials on apartheid, environmentalism and the links between the environment and multi-scalar social justice are woven throughout my work and my life. Muna had moved to Durban when I returned to South Africa for graduate school in KwaZulu-Natal in 2004, and happened to be living in Cape Town when I moved there in 2010. While I had not intended to follow a mentor around the country, we were both grateful that life kept putting us back into orbit. Sometimes the cosmos feels like it has more reason that I am generally prepared to admit. Muna got inexplicably sick in Cape Town, and I was the pseudo-daughter who facilitated getting the western medical attention that extended his life by several years. During my first real effort to get this text into print, in the northern summer of 2019, Muna passed away. Muna was never a big fan of the academy – too conservative and slow – but was always a supporter of me, and I know he would be proud of the text.

My long journey here has been shaped by Garth Myers since roughly the same moment. He was my introduction to geography, African studies, African cities and political ecology during my time as an undergraduate at the University of Kansas. Garth put a pre-print copy of his book *Verandas of Power* (2003) in the library for his students to read, and one of my strongest memories from University of Kansas is the feeling of awe while sitting with a not-yet-published book in my hands. Garth's urgings sent to me to study *in* Africa rather than *about* Africa for graduate school, he served on my PhD committee at Clark, and as a mentor for nearly two decades. As is surely apparent throughout the text, his scholarship has deeply shaped by approach to research and writing, and I also thank him for his thoughtful engagement with a draft of this book. I am also grateful to the Urban Studies Foundation for providing a fellowship to work with Garth in 2014, the first time I intended to put these thoughts in a book form. It turned out I wrote little that was useful in the early months of my son Malcolm's life (he had infant GERD and slept in 2 hour cycles, making us both under-rested and quick-tempered). I thank Garth for not making too much of my zombie-like state at the time.

I owe much to those I worked with at and through the Centre for Environment and Development at the University of KwaZulu-Natal: Rob Fincham, Jenny Clarence-Fincham, Philippa and Jon McCosh, Allison Goebel and Marc Epprecht provided academic support and much more as South Africa became home. Nyambe Nyambe and many of the students I worked with (including Cyprian Alokwu, Pascal Karemera, Nobuntu Gxaba, and Siyabonga Zondi) also helped me to better understand the strengths and weaknesses of the academy broadly as well as in terms of meeting their personal and professional needs.

At the University of Cape Town, Sue Parnell provided the opportunity and support for me to find my way into scholarly writings on African cities and the southern urban critique. Being surrounded by urban scholars at the African Centre for Cities provided a catalyst and crucible for my thinking. My thanks in particular to Jane Battersby, Mercy Brown-Luthango, Shari Daya and Clare Herrick for co-thinking during my postdoctoral work.

I will always be indebted to the Department of Geography, Geoinformatics and Meteorology at the University of Pretoria, where I got my first academic job and the opportunity to stay in South Africa. Nerhene Davis, Paul Sumner, Liesel Dyson and Dan Darkey deserve particular recognition. Leaving the University of Pretoria remains the hardest decision I have ever made, and I deeply appreciated your support through the associated personal/professional struggle.

I met Henrik Ernstson and Jonathan Silver while at University of Cape Town, and it was here that the Situated Urban Political Ecology collective was born. Our personal and professional solidarity has shaped the collective and my own individual identity for the better. I look forward to decades more. Additional collaborators whose thoughts have shaped the ideas presented here, across multiple projects and continents, include Shuaib Lwasa, Tyler McCreary, Gray Maguire, Nate Millington, David Nilsson, Laura Riddering, Kathleen Stokes and Erik Swyngedouw.

At the University of Oklahoma, I am grateful to Kirsten de Beurs and Scott Greene for supporting my work. My graduate students including Anesu Makina, Doug Allen, Lené Le Roux and Aditi Singh have also patiently put up with my quirks, moves, children and high expectations. Working through these ideas together has made them stronger and, I hope, readable for graduate students. The rooibos tea Anesu brought back from South Africa was a constant companion during the writing of this book.

My PhD advisor Jim Murphy has supported the twists and turns in my thinking and career in various ways. The intellectual scaffolding of transitions continues to underpin and shape my work, even when the citations do not directly point there. I would not have persisted through my doctoral studies without the hijinks of the 2007 pirates (Arijit, you are missed), nor come to appreciate urban ecology without Dianne Rocheleau and the unfailingly serendipitous events that arose in the New York City trips.

Not least of which was seeing Joe Pierce "in place." We share a love of words and an appreciation for finding the right ones, but words will never get us right. You are the first and last audience in mind as I write, and this book is clearer and stronger for knowing you will be its most careful reader.

Malcolm, Rowan: you are the reasons this book did not happen years ago, but I would not have it any other way. (And I hope the ideas in it are stronger for having waited.) In the long northern summer of 2019 when this book really happened, I was more playful, patient and present when the writing went well, and less so when an idea refused to stay at work. The fish we acquired during this time were no substitute. I had not fully understood until now why authors thank their families for bearing with them while writing a book, and how different it is from the more punctuated process of article writing. Your giggles and gazes brought me out of my thoughts and back to the world and were a reminder of why all this matters in the end.

Funding to support writing and the waste work was received from the Economic and Social Research Council–Department for International Development joint fund for poverty alleviation research ES/M009408/1 for the project "Turning Livelihoods to Rubbish? Assessing the Impacts of Formalization and Technologization of Waste Management on the Urban Poor." Funding to support writing and the infrastructure work in Kampala was received from Vetenskapsrådet 2015-03543 for the project "Heterogenous Infrastructures of Cities in Uganda Project: Thinking Infrastructure with the South."

Lightly edited parts of the Introduction, Chapters 1, 2 and 6 have been published as:
Lawhon, M. and Truelove, Y. (2020). Disambiguating the southern urban critique: Propositions, pathways and possibilities for a more global urban studies. *Urban Studies*, 57(1), 3–20.

Lightly edited parts of Chapter 5 have been published as:
Lawhon, M. and Le Roux, L. (2019) Southern urbanism or a world of cities? Modes of enacting a more global urban geography in textbooks, teaching and research. *Urban Geography*, 40(9), 1251–69.

Introduction

Things Fall Apart (Achebe, 1958) is surely one of the most read and best-known African novels. Aptly titled, the work describes the disastrous disruptions of the colonial encounter in the fictitious village of Umuofia in rural Nigeria. I start this monograph instead with the less well-known second book in Achebe's "The African Trilogy." *No Longer at Ease* (Achebe, 1960) takes place at a later moment, something more akin to Achebe's present. The protagonist is a young man, grandson of the infamous Okonkwo of *Things Fall Apart,* raised in the same village of Umuofia. *No Longer at Ease* follows Obi Okonkwo's struggles to create a new life in Lagos after returning from university studies in England.

The political critique of *Things Fall Apart* is rather straightforward. Achebe provided a counter-narrative to prevailing literary representations of Africans as "in need of colonial rescue" through his presentation of Umuofia as a reasonable moral universe predating colonialism. Achebe's sequel, however, grapples with a more ambiguous set of concerns. The title, *No Longer at Ease,* is a reference to a British poem, a line from TS Eliot's "The Journey of the Magi." Achebe's characters do not yearn for a precolonial, unobtainable past, but instead wrestle with what to do in a moment in which the black and white morality that shaped Achebe's narrative of the colonial encounter has become much more gray. There is less of the violence, more of the personal struggle between Europe and Africa and an awareness that this "in between" position is the source of much unease. Obi appreciated his British education (including switching his major from law to English, a thinly veiled parallel to Achebe's own switch from medicine to English), even if he also often missed Nigeria. He is "uneasy" with a friend's anti-Englishness, and struggles with the family expectations and social norms regarding whom he ought to marry. This is not a story of a struggle to be more European, though; Obi's father is negatively portrayed as excessively and uncritically adopting of European ways. In the lack of resolution at the end of the novel, we may read a suggestion that Achebe was unable to find a way to reconcile this feeling of being uneasy. Interestingly, Achebe and/or his audiences may well have found this uneasiness and in-between-ness too much to explore further: Achebe returned to the earlier colonial moment and its clearer moral universe in the concluding novel of his trilogy.

No Longer at Ease does not undercut the colonial critique of the classic *Things Fall Apart,* but instead grapples with a different question. Its characters struggle through two juxtaposed cultures, neither of which appears to be satisfactory in themselves. Obi's struggles are internal, but also create uneasy relationships with others. This process, difficult as it is, continues to be relevant to our present, for the questions we might leave *No Longer at Ease* with focus on what to do, what to create, how to uneasily work at the juncture of the easier (but in this novel, both unsatisfactory) dichotomies of Europe and Africa.

I use this novel as a starting provocation through which to think about what it means to work through both northern and southern theory as well as northern and southern cases. No doubt it is easier to embrace a side than to wrestle with ambiguity, whether that side is African or European, and many continue to do so in and outside the academy. Of course, the binary of Europe and Africa, or north and south, never was and remains unsatisfactory, a point I return to further below. But the heart of Achebe's concerns, working between two quite different ways of being and understanding the world usefully draws attention to the unease of being in between. This book is about how we might seek to inhabit uneasy intellectual positions and what we might learn and unlearn in the process. Most importantly, it is also about what can emerge from this process: an ability to theorize new thoughts that draw on but cannot be subsumed within more established ways of thinking.

Towards a better understanding of southern cities

My work is motivated by and seeks to contribute to the growing scholarly interest in cities in the global south and ongoing struggles over how to study them (e.g., Robinson, 2002, 2006; Roy, 2009; Watson, 2009; Roy and Ong, 2011; Parnell and Oldfield, 2014; Robinson and Roy, 2016; Lawhon and Truelove, 2020). The majority of urban residents already live in the global south (more than half of the world's urban residents live in Asia alone) and urbanization rates in southern cities outpace their northern counterparts. For decades, scholars have argued that urbanization in the global south needs to be better understood, and paired this with a concern that we cannot simply deploy established urban theories and tools to southern cities (e.g., McGee, 1971, 1991; Myers, 1992; Sanders, 1992; Slater, 2004). With Yaffa Truelove, I term this long-standing argument that has recently reinvigorated urban scholarship about global south cities "the southern urban critique" (see Lawhon and Truelove, 2020 and Chapter 2 for more detail on its roots and various iterations).

While there is a longer history to this concern, something has changed in the past decade or so in urban studies. Robinson's engagements (2004, 2005, 2006) were instrumental in instigating a wider, overdue conversation about the relationships between cities in the north and south and urban theory. Awareness and acceptance of the significance of cities in the global south is increasingly evident in urban studies. The works of some of its most well-known scholars are frequently cited, plenary and keynote talks at major disciplinary conferences have been given and students are being trained and contributing to growing scholarly conversations (e.g., Robinson, 2011; Roy, 2011; Pieterse, 2013; Lawhon *et al.*, 2018c). Urban journals include editorial boards with southern scholars, statements urging submissions about and from the south (e.g., Leitner, 2008; Seekings and Keil, 2009) and funding opportunities for scholars from the global south (e.g., Urban Studies Foundation, undated). A 2013 manifesto consolidated many recent arguments (Sheppard *et al.*, 2013), evidence of a common cause amongst a growing collective of southern and south-aligned urbanists.

Despite a growing awareness of the theoretical and empirical importance of understanding southern cities, as a scholarly community we remain uncertain and ambivalent about how to proceed. Creating north-south and south-south research networks, providing translation services, citing southern scholars and increasing comparison have been offered by scholars as avenues through which to reshape the flows of knowledge (e.g., Robinson, 2011; Hart, 2018; Collyer *et al.*, 2019). These tactics may well increase the presence of southern voices and southern empirics in the "international" literature. In themselves, they are worthy goals for redressing injustices within international academic hierarchies (e.g., Patel, 2006; Ferenčuhová and Gentile, 2016). Esson *et al.* (2017) argue against a focus on knowledge production and instead emphasize the "structures, institutions and praxis" of the academy and the voices of marginalized scholars. While I believe these concerns are important and have written about strategies for creating more just institutions (Lawhon, 2018a), the connection between identity and the production of insights from beyond the global north theoretical canon is tenuous (see Nagar, 2014 on the need to do more than identify identities in our scholarship). Sanders' (1992: 206) trenchant argument may be several decades old, but continues to have salience today: "[T]here are numerous ways in which Western social science interpretations of Africa have introduced distortions not only for Westerners but also for Africans who rely on Western interpretative models to understand their circumstances." (I engage with her arguments in more detail in Chapter 1). Colonialism and racism were and are powerful forces in the world, and their impacts continue to be felt in formal and informal educational systems across the global north and south. Beliefs in the promise of modernity and development remain prevalent, and hard to disrupt, particularly in the absence of compelling alternatives (Ferguson, 1999). In this context, it seems that identity might well matter, but the relationship between identity and counter-hegemonic knowledge production has been vaguely alluded to in the scholarly literature, and particularly in urban studies.

My contention is that the strongest and most compelling version of the southern urban critique goes beyond a need to better include southern voices and southern data (again, see Chapter 1). It is that we need to rework the process through which we engage with southern cities and the ways we write about them. No doubt more data and voices are easier: voices and data are easier to measure and easier to see, even if bringing them into urban studies remains difficult to accomplish in practice. The difficulty of stepping outside what we know is well-established: even without critical theoretical concerns about modernity, development, race and power, Kuhn's (1962) classic work on scientific revolutions shows us how difficult it is to generate insights outside of existing modes of explanation. If we could stand outside of ourselves, we might simply look to comparison and dialectics as modes through which to generate new ideas from two incomplete ones. But, as long argued by poststructural, feminist and postcolonial scholars,[1] we are all too embedded in our world for such objectivity.

My focus in this monograph, then, is the process of postcolonial theory-making. The decades of writing following Spivak's explication of and ruminations on unlearning have profoundly shaped scholarly engagements with reflexivity, yet not left us with a particularly clear and succinct iteration of the process (e.g., McEwan, 2003; Kapoor, 2004; Lawhon *et al.*, 2016). Equally, Bhabha's work on hybridity has productively worked to de-essentialize our ideas of identity and culture. Many have pointed out, however, that Bhabha's ideas may be appropriated to justify decontextualization and erasure of culture and identity (e.g., Mitchell, 1997; Easthope, 1998; Werbner and Modood, 2015). In seeking to work through my own experiences of hybridity and unlearning, I do not mean to suggest we need a formula or a step-wise methodology for postcolonialism, for reworking the process through which we produce theory is not easy to do nor easy to see nor translatable into concise research methods. I mean to say, we need more accounts of individual, situated and relational processes of unlearning from which to generate deeper insights into how to facilitate and recognize this process. And maybe, make more space within the academy for it. This book provides one account of my academic journey, situated within the wider literature and attentive to the process of unlearning and learning anew.

This book, therefore, is not a state of the art of urban knowledge about the global south (see instead Miraftab and Kudva, 2014; Parnell and Oldfield, 2014). It is not a methods book for how to undertake research, even of a postcolonial kind. It is an examination of my own research trajectory with a particular emphasis on the process of generating new insights. This book is deeply personal, although I try to limit my navel-gazing in the text itself and focus on processes and lines of argumentation and their wider relevance. I seek to contextualize the contributions I have made to the scholarly literature by giving them longer histories, showing how they are rooted in places, people and encounters as well as postcolonial analysis. The point here is not, however, to call attention to my research outputs but to attend to the modes of their production. South African cities and African scholarly literature are at the heart of the text, and I make no apologies for this, but do also identify where other sites and literatures have shaped my thinking. I am trained as a geographer, and while I reach beyond urban geography, the work too shows this bias.

As I have been writing, I have also been working to train three doctoral students/candidates from the global south. This book is written in particular to and for them (and parts with two of them!) I hope it is of use for their many peers across the north and south seeking to navigate what is a high energy, high stakes and often rather muddy set of arguments. My hope is that both the personal and professional narratives, as well as what I hope is (reasonably!) straightforward language, make it easier for them to productively situate themselves in, and engage with, the ongoing dialogues around southern cities.

From learning to unlearn to learning anew

Distilling key processes in hindsight is, no doubt, a tricky practice. I work to be cautious in my presentation here, for the story I tell throughout this monograph is clearer and cleaner than the imperfect and wandering road of practice. Nonetheless, my effort here is not to write an auto-ethnography or testimonial, but to connect experiences and lines of argumentation with particular insights. My hope is that working between auto-ethnography and a more typical result-oriented research monograph helps the reader to see what underpins the work while also working towards more generalizable processes. I draw on and work through post-colonial literatures on hybridity and unlearning in order to attend to the ways in which *juxtaposition* can induce the hard work of *being uneasy,* contributing to *unlearning* a (comfortable) gaze and enabling researchers to *learn anew* about (un)familiar places. Cautious that this quartet might turn into a formulaic research methodology, I show throughout this volume that this is an iterative, ongoing process of reflexivity rather than a recipe for postcolonial method.

First, *juxtaposition.* Juxtaposition[2] has been central to my own scholarly process, for putting two things next to each other helps one to see similarities and differences. Juxtaposition, as I argue in more detail in several parts of this book, is useful but in itself an insufficient step in the process of postcolonial theory making. For the juxtaposition of difference is long-standing and typically has provoked polarized responses: "liberal" celebrations of diversity as beautiful, cosmopolitan and nonthreatening and "conservative" hierarchies of difference and/or defences against the despoilation of cultures (see Ahmed, 2012). The colonial encounter was a form of juxtaposition, and study abroad and voluntourism activities continue this process of visiting a broadly defined "elsewhere." My own experiences as a mentor for both study abroad and voluntourists, as well as the wider literature, suggest that without careful attention such experiences may well reinforce negative perceptions rather than prompt their deconstruction (e.g., McLennan, 2014; Loiseau *et al.*, 2016; see also Chapter 5 on teaching difference). Juxtaposition alone does not require us to work through our own gaze.

Unease. Recognizing that two things are irreconcilable can produce unease.[3] Often, our experiences stop here, causing us to retreat to safer ground. Much of EuroAmerican scholarship has long taken ideas from the south to be both irreconcilable with EuroAmerican knowledge and (therefore) wrong. Ideas from "elsewhere" are recorded in anthropological accounts, but cordoned off from "universal" theory (Hountondji, 1996; Connell, 2007). This tendency to separate bodies of thought has a parallel in some the Africanist scholarship arguing for a distinctive African ontology as well as social and political philosophy. Philosopher, poet and Senegalese President Leopold Senghor is probably the best known for articulating this position as "Negritude" (Senghor, 1966), and such arguments have recently regained prominence in global academic discourse. I reject both northern and southern iterations of this position.

In doing so, I draw on southern and black scholars who reject totalizing theory as well as racial or ethnic essentialism (Hountondji, 1996; Connell, 2007; Nagar, 2014). As Gilroy (2019) has recently argued, "we should be sceptical about the seductions of the ontological turn recently promoted in the study of race politics. It has become disastrously complicated by prospective nostalgia for the easy, essentialist approaches that were dominant when assertive cultural nationalism ruled the roost." This is not to say that there is one right way to be in the world, to deny different knowledge traditions or even that there is one right way to undertake research. It is a rejection of the idea that the process of reasoning and/or being ought to be different for different races, nationalities or cultures, of a relativism that lets us leave what is irreconcilable alone and chalk it up to difference. Here, I equally reject northern and southern intellectual isolationism as well as any suggestion that either already has all the answers (or an exclusive foundation from which to develop them). I am interested, instead, in the refusal to retreat to what is easy: I am interested in this unease, and in working through it towards more rigorous answers.

This next point was initially absent from the text, but the more I wrote about my own process in this book, the more I realized the importance of considering unease and pre-existing relationships. As was apparent for Obi in *No Longer at Ease,* working in the in-between space had impacts on his relationships with others: Obi was uneasy both with his Anglophilic father and the Anglophobia expressed by friends. Obi's in-between-ness was also the source of much angst for his family and friends, particularly as he grappled to identify and explain it. While I largely try to leave innocent others outside my text, I do at times point towards ways that this wider journey has created relational unease and think about the consequences of this for our scholarship and willingness to undertake uneasy work.

Unlearning means turning unease into an examination of the construction of individual and social gazes. Unlearning requires us to be reflexive about the production of hegemonic, counter-hegemonic and individual views and values, and specifically, the difficult task of identifying the assumptions that underpin why we see what we see. It is also about the interplay between views, values and research. In my experience, in the long run, unlearning has been helpful for mitigating my own intellectual unease, quelling contradictions in a search for a more compelling and consistent gaze, narrative and politics (although it has been less satisfactory for mitigating unease produced by having different ideas from others, what I describe as relational unease). The call for unlearning through reflexivity is widely present in the literature (e.g., Danius *et al.*, 1993; Spivak, 1999, see also in urban studies Porter, 2004, 2016; Chattopadhyay, 2012; Lawhon *et al.*, 2016), but I have often found it difficult, particularly when analyzing this literature with graduate students, to think through what brings a person to this point. How do we identify what ought we be reflexive about? What prompts this reflexivity? In lengthy conversations across multiple continents, my sense is that my experiences that juxtaposition and unease have instigated many searching conversations is not atypical, even if rarely present in the text version of our research. Adding juxta-position and unease enables us to lengthen the process of unlearning, hopefully providing more clarity as to where we might start, as well as a reminder of what to return to in this iterative process.

It is after this unlearning that postcolonial *learning anew* often happens. In much of the work that I detail below, this has meant developing different starting points for research. This includes rethinking what is included and excluded in the term "the urban," and from here differently understanding what the urban is and might be (Chapters 3 and 4). It includes rethinking how we teach about cities in the global south, and from here, thinking anew about the teaching of urban studies (Chapters 5 and 6). It includes separating our understanding of how cities work from our assumptions about what they ought to be (Chapter 9). In keeping with my argument about the unity of knowledge, these insights are generated from my work in southern cases, and often have greatest relevance to southern cities. In many instances their relevance spans beyond the south, although this is not the focal point of my broader scholarly agenda or line of argumentation. The type of learning anew I describe in this book, then, is a product of reworking the way we see a problem, being open to redefining it and redressing it in often unexpected ways.

Focusing on the process of theory-making, and the juxtapositioning, unease and unlearning that enables learning anew, helps us to navigate the ongoing con-flations between *who* can respond to the southern urban critique, *what it takes* to do this work and *how scholars can recognize whether this work has been done*. Each of these is a slightly different question, but entails a substantively differ-ent diagnosis of the problem(s) with contemporary urban studies. I argue that scholars ought to reject the essentialism inherent in the suggestion that *being from* a place grants a particularly postcolonial gaze. Creating a more inclusive academy is, in itself, a task worthy of our attention (see, e.g., Lawhon, 2018a), and it is well-established in the social sciences that the identity of a researcher shapes participant responses. Neither of these points, however, excludes anyone from having the potential to generate accurate and useful insights (see Larner, 1995; Jankie, 2004; Raghuram and Madge, 2006; Nagar, 2014). I draw here, for example, on Rocheleau and Edmunds (1997) (drawing on Haraway, 1991; see also Nagar, 2014) to think about *affinity* rather than *identity*. I build on this argument to articulate in more detail what affinity might entail and how it might be produced, mindful that something akin to affinity does not imply agreement.

This argument, then, might help us towards explaining why identity often enters conversations about postcolonial knowledge. In the world we have, being from the global south and/or having a non-white racial identity is not irrelevant. This is not because of any essence or birthright. Instead, in the world we have, the "double consciousness" articulated long ago by WEB Du Bois continues to be impactful (1990 [1903]; Bruce, 1992). Double consciousness means living in a world of juxtaposition, and this double consciousness produces plenty of unease (Du Bois, 1990 [1903]; Fanon, 1952; Oluo, 2018). I recognize my inability to do justice to the extensive literature and advances since Du Bois (but do briefly point to geographers who have usefully contributed to articulating spaces of respite that seek to mitigate this unease, e.g., Bledsoe, 2017). My point is that while many people have to seek out circumstances through which juxtaposition might produce unease, this is not the case for most people in the world.

And yet. It is worth repeating that unease does not *necessarily* mean unlearning and theorizing anew, which is why we must attend to much more than unease. Throughout this book I work to think through how unease might, at times, usefully prompt unlearning and learning anew. Returning to Achebe, our search here is the search for how to live in a world shaped by Europe and Africa, living a life that is entangled with and inseparable from multiple locations and cultures which are themselves relationally produced (Massey, 2005). It is the question of how we might hold on to what is useful/good/right from both Europe and Africa that makes Obi no longer at ease. Earlier in Obi's life, at earlier moments in our global history, it was easier to be at ease in a singular gaze. But even as colonialism and its successors have bolstered the hegemony of the north, juxtaposition is increasingly inescapable in our interconnected, cosmopolitan world. The reactionary forces in contemporary politics make it clear that many are feeling threatened by the fact that it is increasingly untenable and impractical to build walls, material or otherwise. I work from the belief that we do not already have the answers about what to do, but that learning anew can help us to better understand our world and the world we might want to create.

From Kansas to KwaZulu-Natal

Reflections on race and nationality are not the only ways through which we come to struggle with our gaze, although these are typically central in postcolonial scholarship. I trace my own awareness of the irreconcilableness of different worldviews to the clash of ideologies produced in my Catholic upbringing. In no way do I mean for this to be of equivalence with other forms of identity that have received greater attention in postcolonial studies; my point here is that unlearning can have many different prompts, and that we might learn from embracing, not discouraging, these different journeys. Recognizing that people can come to unlearning through different journeys usefully opens postcolonial theory-making to a wider array of participants and perspectives, and may well make it easier for us to imagine cultivating postcolonial sensibility across urban studies, a point I return to repeatedly below.

I went to Catholic schools from kindergarten through high school, and as much as one imagines growing up to be anything at ages ten and fourteen, I imagined a vocational life within the Catholic church. But being a priest was not allowed. The gender narratives I adopted from mainstream US American culture – that my gender should not constrain my professional choices – were incompatible with Catholic teachings. I could not ignore the juxtaposition of these two perspectives, and being excluded from the most visible role in the Catholic community made me deeply uneasy. I searched for answers within Catholic theology (troubling no shortage of teachers and priests with my enquiries). I spent much of my early high school life in an effort to come to peace with the answers provided within the system. I *believed* I would find answers within the system. I *wanted* to find answers within the system. But my convictions about gender equality were ultimately incompatible, irreconcilable and stood on firmer ground. (This, of course, raises questions in the unlearning process about how one identifies these bedrock principles and their social construction; while I do not have the answer, I reflect more on this struggle in Chapter 9.)

The oldest story I knew, the one that shaped my life, my relationships, my imagination of who I was and the future I wanted, crumbled. It was unexpected, not what I had sought, hoped for or wanted. I was heartbroken. Lost. It did not help that my struggle was not only internal: it turned relationships upside-down. My family. My father urged me to tell white lies to my grandmother for years in hopes that my defection was a temporary phase. And at that stage, a product of Catholic education, every friend I had was Catholic too.

I found my way through that phase, a story for a different book. What matters here is twofold. The first is that it generated a lifelong skepticism of authority and compulsion towards unlearning assumptions. The second is that I was equally unable to accept mainstream US American cultural discourse. It too was limited, and failed to address much of what I had found compelling in Catholicism, including a sense of collective purpose and responsibility to others.

This experience left me with a lot of uncertainty and curiosity, and prompted a search for more juxtapositions through which to rebuild a worldview, to learn anew. Through another series of events, again for a different book, South Africa became the place where I would continue this process of juxtaposition, unease, unlearning and learning anew. I went for the first time in 2002 for a six-month voluntourism experience, taking a semester off from my junior year at the University of Kansas. (Yes, I am critical of voluntourism and even at the time was aware I would learn more than I could give.) At that time, I had never left the United States and had lived my whole life within a hundred miles of where I was born. To say this was an uneasy choice, a choice to be uneasy, is an understatement. It furthered the rift with my family. This was not only about spatial distance; although we never used such words, I believe they understood I was continuing the process of unlearning the worldview we had previously shared.

To quickly round off the outlines of the biography: I returned to Kansas and finished my undergraduate degree in 2003. I then went back to South Africa to start a master's program at the University of KwaZulu-Natal's Pietermaritzburg campus, where I eventually spent several more years as a lecturer and research assistant. Back to the United States in 2007 to start a PhD program, I returned to South Africa in 2010 for dissertation fieldwork. During that time and the years that followed, I did research. I taught. I wrote. I walked. I walked a lot. I did my postdoc at the University of Cape Town and got my first academic job at the University of Pretoria. I ran by the ocean, made friends, learned pieces of more languages, bought my first home, took minibus taxis and eventually got a scooter, had it stolen, got another, had surgery, joined protests, co-organized protests and spent more time than I expected in hospital.

My hospital stays included having my first child, born just a kilometre away from the Union Buildings. But unforeseen and immediate changes in South Africa's visa laws, and additional difficulties with the new baby meant I made the easier choice: I left. I took a job at Florida State University, and eventually moved to the University of Oklahoma where my family is in closer proximity.

I draw attention to this decision about going or staying because, while my argument in this book is that unease can be productive, it equally worth emphasizing that too much difficulty can hinder thinking and writing. There are surely limitations to the idea that we all need a room of our own to write: many contributions have been written under much more constrained conditions. Alice Walker's (2004) argument is not quite oppositional to the classic feminist argument of Virginia Woolfe in its essence, but is meant to push us to think beyond a physical room and towards a wider sense of what it means to have the "room" to write.[4]

We all must find our balance of the different kinds of unease and room. I made the decision that overall my scholarly contributions would be better with fewer everyday stresses. Moving to the United States meant more help from my family, more money (my take-home salary tripled), and more everyday at ease. I worried less about remembering to lock a door and turning my back on my groceries (for all the hype and reasonable criticism of it, crime in South Africa is a real concern and was a regular part of my life). While everyday stresses in South Africa for me were manageable – even productive – as an individual, their impacts multiplied when I became responsible for more than just myself and, sleep-deprived, struggled to remember let alone think and write anything.

No clock that can tell whether this is enough to claim knowledge situated in a place. As with race and nationality, time can matter, but in itself is insufficient. Instead, I found and find more utility attending to the emotional and intellectual process of change that happened over many years. I learned to revel in what was uneasy. This was not about seeking adrenaline, of intentionally putting oneself physically or emotionally at risk, but about being intellectually and emotionally not at ease as I sought to reconcile the different ways of making sense of what I saw, read and lived.

Most Americans like myself who grew up surrounded by whiteness struggle to talk with people who, at first glance, appear different than them. It takes work to transcend this. I was absolutely uneasy in conversations with people of other races and from other places for many years. A black child once told me my laughter sounded nervous, his word. Uneasiness, my word, made it hard to find words, any at all. And when found, they were often not quite right, and even wrong. My co-voluntourists in 2002 were all much more worldly than I, and it showed. I was still wrestling with basic things like "black people are harder to see at night" and "do you cross the road when walking past someone who looks [fill in the blank]." (An easy admission? No. I share as context, and in hopes that this text might also reach people at various points in their journeys. I was a late starter; it is never too late to start.)

In South Africa, being an outsider meant that I was permitted to engage across racial groups in a way that is particularly difficult for white South Africans.

Several black South Africans said precisely this on different occasions in different places across the years. I was given opportunities to fail, flop, offend and try again. Muna, whom I wrote about in the acknowledgments, put up with me. In general Muna was not a fan of the United States, but in our many years of working together, he came to understand that I had come first to learn or, in the vocabulary I later learned, to both learn and unlearn.

Towards the end of my stay in 2002, I moved to Dlamini, Soweto, just across the road from Regina Mundi Catholic Church, where I lived with a family who both put me up and put up with me for a month. During this time, I came to not just not be uneasy when surrounded by difference: I found myself at my calmest when squeezed in a taxi full of other passengers, most at ease when actually touching other people. Knowing the relevant hand signals, being able to confidently pass money forwards and back, pronouncing "s[h]o[r]t left" with the right inflection at the right volume such that it was actually understood the first time. I was different, and everyone in the taxi knew it. In the taxis to Soweto other passengers sometimes politely asked if I was in the right place, headed to the right place. But when I gave the right names, we shared a sense of togetherness. Affinity. Every time I re-read this paragraph from my office in Oklahoma my heart hurts a little with a feeling I can only identify as a homesickness of sorts.

For most of the time I lived in South Africa, I had a budget of about $1k a month. This put me in the top 10% of households, and for most of this time, I had no dependents, no family members I needed to send money to and no one needing caring labour from me. I was lucky that my own health issues never overlapped with teaching or research commitments and meant I could take time off work without consequence. While some things are less expensive in South Africa, on the whole, the cost of living was not notably cheaper than in Kansas. My framing has largely been that it is easier to live inexpensively in South Africa: taxis come more often and go more places than middle US American buses, most people have wardrobes with fewer and less-expensive clothes and so on. I felt reasonably comfortable in a lifestyle where I took mass transit, lived in backyard "granny flats" (typically converted maid's quarters) and travelled inexpensively (barring flying back to Kansas roughly once a year); this material simplicity was never uneasy for me. In Pietermaritzburg, it meant I lived materially most like the black students from across Africa, which facilitated social relationships; many of the South Africans and all of the other foreign white students had cars and their own separate social circles, and lived further from campus. As above, I do not claim that this experience made me similar; I remained privileged, different. But it did produce an affinity and opportunities for interaction that most scholars, particularly northern ones, often struggle to develop.

I also grew able to be at ease with making others uneasy. Most memorable here is my experience taking my two-year-old black South African neighbour grocery shopping. For part of my time in Pietermaritzburg, I lived in a block of flats with mostly black South Africans residents, and a couple of the young children would regularly come visit. I never properly met their parents; they just waved and called me "gogo," the widely recognized isiZulu term for "grandmother" (implicit in this moniker is that grandmothers often care for children who are not their own). One day, one of the kids asked to accompany me to "the shops" across the way, and with his parents' consent, we went. Older white customers stared. Some were openly hostile, shaking their heads at me, assuming that the child was mine. The responses from strangers were much more aggressive than when I walked around with black African friends, even male ones. I am confident the child did not make much of this, as it became a habit for us to go together that he clearly quite enjoyed. I both came to take small pleasure in being a disruption and much more viscerally aware of how hurtful it could be to be treated so harshly while simply walking through the world with a child I cared about. I am mindful this was both constructed and optional for me, and again in no way mean to equate this with what others experience. Affinity, not equivalence.

And so: can I write about postcolonial thought and African cities in scholarly publications? Sure, I have, and many publish from much less than this. Do I see the world as a South African? No, surely I do not. But nor do I quite see it as a US American. I have doubts as to whether either of those framings makes much sense, particularly in the world we have today. Instead, I am interested in the conceptual, temporal and spatial juxtapositions, working with unease, excavating assumptions and then trying to put back together a new way of explaining the world and what it might be. This process has been generative of thoughts that are products of inter-connected, multi-sited places, people and encounters. My insights are a product largely of me, being a changing and learning me, thinking in and through southern cases, although I would not call them "southern theory."

The outcome of various stories that brought me to the moment when I am pulling these thoughts into text was surely not intentionally cultivated: I did not seek to be a hybrid developing affinity and unlearning in order to be a better equipped scholar. Nor do I suggest that it is reasonable to expect the typical gradu-ate student to spend a decade working through such processes before publishing or being hired, or that the processes outlined above are the only way to conduct good research. Some can excavate old texts and see in them, or in reading across them, what others have not. Some derive new insights from empirics. Insights gen-erated through postcolonial unlearning and learning anew will likely be advanced through both reading of texts and developing new empirics. My urging here is for us to be clearer about the process of postcolonial learning as one way to think through a southern city and develop new insights, as well as to increase our ability to recognize and valorize both the process and the insights generated through it.

Being uneasy, unlearning and learning anew is hard, often underacknowledged, intellectual work. It is emotionally difficult. It works in unpredictable ways on unpredictable timelines. We as people and as scholars often shy away from uneasy tasks and experiences, and return to safer, more solid ground from which we can speak with confidence. Postcolonial theory-making is hard political work too. It would be easier to count simpler metrics, and for progressive academics it might equally be easier to valorize non-western worldviews and explanations. Yet my argument here is that we need more not being at ease. In a world full of answers, none of which quite works, juxtapositioning, unease and unlearning can be steps towards thinking anew. And thinking anew can open new possibilities, enabling a more just world; this more just world is, at the end of the day, what motivates my search and research.

Outline of the book

In the next chapter, I turn to the academic literature to situate my argument in urban studies. I identify the southern urban critique as an intellectual and political move within urban studies that interrogates the construction of knowledge about cities in the global south (and distinguish this from the more general use of postcolonial theory to study cities). I then work through early iterations of this critique from the 1990s, as postcolonial theory became established in the wider field of geography. The chapter concludes with a brief note on Robinson's mid-2000s engagements with urban theory which were instrumental in instigating the more sustained engagement with the southern urban critique in urban studies that continues into the present.

While Chapter 1 sketches a longer history, Chapter 2 provides a framework through which to understand ongoing questions in contemporary global south urban studies. The chapter draws on an article written with Yaffa Truelove and published in *Urban Studies*. The southern urban critique has troubled the core of urban studies, but there is much uncertainty as to the precise arguments of authors in this wider community and what they want to see change. Chapter 2 focuses on different foundations for this critique: empirical difference, different intellectual traditions and the situatedness of all knowledge. We argue that the most rigorous version of the southern urban critique calls us to attend to the process of theory-making for theories generated from cities across the north and south. This is not incompatible with a deeper awareness of empirical difference or more careful engagement with other intellectual traditions, but provides a different root for the critique and thus how to address it. This argument is no simple call for a global urban studies, but an urging for all urban scholars to engage with postcolonial concerns about the situatedness of knowledge and the limitations of universalizing theory. I take up this version of the southern urban critique in the remainder of the book.

Chapter 3 engages with the very foundation of urban studies through an enquiry into the urban question: what is a city? This chapter provides an analysis of the ways in which scholars of African cities have deployed northern answers to the urban question. I draw here on Robinson's (2006) assertion that we can look to the 1960s anthropological literature as an inspiration for a more cosmopolitan urban studies. I agree that there is utility here (and I hope it is clear that the wider arc of my work is deeply indebted to Robinson's scholarship): the 1960s literature does trouble the universalizing teleology embedded in urban theoretical assertions of the time by showing that migration to a city does not necessarily result in the adoption of a particular type of urbanity. However, the precise argument deployed by various anthropologists should give us pause: the authors build their argument by identifying places that do not adopt urban lifestyles as "un-urban," or rural-in-the-urban. In other words, this literature problematically deploys northern definitions of the urban rather than opening up the urban question to African empirics. In doing so, it also reinforces the conflation between urban, European and modern that pervaded the colonial encounter in Africa and beyond and, I argue, continues to have relevance today.

Chapter 4 develops this argument about the urban question and the conflation of urban, European and modern through a contemporary empirical case. It also explicates the relevance of such questions for contemporary urban scholars, states and residents. The chapter builds on a series of photo-elicitation interviews conducted by my research assistant and now PhD candidate Anesu Makina in several different suburbs in South Africa in 2017. These interviews generated much more dissensus than either of us had expected. Disagreeing perspectives appeared to be shaped by race, age and the location of the interview. They showed that the city continues to be narrated by many respondents as being comprised of modern "white" or historically white spaces. This perspective, however, was far from uniform. Respondents in the township, for example, were much more likely to declare a more inclusive vocabulary, which we interpret as a desire to be included in the city. We conclude this chapter with thoughts on the implications of these findings for research, policy and everyday experiences in and of the South African city, including the awkwardness of these findings for "city visioning" exercises as well as ongoing critical discussions about the "right to the city."

Chapter 5 then turns to the question of how to teach urban geography in light of the southern urban critique. I draw here on published work conducted with another research assistant and now PhD candidate, Lené Le Roux. We analyzed five contemporary urban geography textbooks and the different modes through which the authors incorporated southern cities and the southern urban critique. We then reflect on the strengths and weaknesses of various approaches. Given the world we live in, we believe there is no perfect solution for incorporating southern cities more deeply into urban geography textbooks: students will understand familiar cities better and most students, whether from the north or south, enter the classroom with colonial modern gazes through which to view southern cities. We conclude with an argument drawn from the wider thrust of this book: urban geography textbooks and courses ought to explicitly address postcolonial concerns with representativity and representation, as well as the limits of existing urban theory and practices of urban theory-making. This may well make both instructors and students uneasy, but by now it is hopefully clear that I think this is a crucial part of the learning process, albeit not the only way to learn.

Chapter 6 takes these different modes of incorporation in urban geography textbooks as a starting point through which to ask wider questions about the ultimate goal of the southern urban critique. I draw here on the same manuscript written with Yaffa Truelove in *Urban Studies* (see Chapter 2) to articulate three different versions developed from our review of the literature. These versions of what it might mean to incorporate the southern urban critique into urban studies are not compatible with each other (although could coexist in a pluralist field). We call these southern urbanism, global urbanism and a postcolonial world of cities.

While this effort to articulate different modes of incorporation helped me (and hopefully the wider field) to develop a deeper understanding of ongoing scholarship on the southern urban critique, it also left me with a sense of being at an impasse. I struggled to quite make sense of where I stood at the end of the paper: aware of the limitations of southern urbanism and global urbanism, but without a clear sense of what precisely a productive relationship between "theory" and a postcolonial world of cities might entail. The *Urban Studies* paper concludes with a vague gesture at this realization and a question as to whether we might benefit from thinking more about what we mean by "theory."

Chapter 7, then, grapples with what it might mean for the southern urban critique to think more carefully about what we mean by "theory." With Lené Le Roux, I construct an argument for more carefully identifying the components of theory as part of the wider conversation on southern cities and their relationship to northern theory. We use the examples of recent debates over urban dispossession and gentrification to show that various terms get used in different ways, conflating empirical objects and processes with causal explanations and normative judgments. This is not meant to be an exhaustive study, but to exemplify how terms are used in specific scholars' works. We use the term "components" loosely here: our goal is not to identify a set of discrete "objects" which then add up to a theory but to point to different ways in which scholars construct arguments about unlearning, provincialization, southern cities and urban theory. (In the background of this argument is ongoing work with Joe Pierce and my PhD students and our collective engagements with "place." Doug Allen, Joe and I published an article that asserts the compatibility of black geographical approaches with relational place-making (Allen *et al.*, 2019). Less evident in the published article is my own set of questions: how might the study and theory of place-making need to change if the sites are black spaces or southern cities? This work is ongoing, and Lené Le Roux, one of my graduate students and research assistants noted above, is carrying this forward in her dissertation work. I raise it here as an example of juxtaposition and unbundling in the global north, as well as a reminder that I too am still unlearning and learning anew as new ideas and places enter my thinking.)

Chapters 8 and 9 then turn to a deeper analysis of what lies behind my own published research in urban studies. I reflect on four different areas of my work to demonstrate different ways in which I have argued for the provincialization of particular terms. I use the term "provincialize" to mean putting something in its place. In other words, provincializing means demonstrating how what is allegedly universal is actually rooted in a particular location (Chakrabarty, 2000). In so doing, the focus of these chapters is not to convince the reader about the need to think about any particular concept from a southern perspective: the previously published work is intended to do so. Here, I seek to step back from the concepts themselves and focus on the structures of the arguments. In highlighting the variety of arguments, I hope to help us to think more carefully about the components of theory being examined in any given work. My point is not to suggest that there are more or less correct arguments in the abstract, but to demonstrate that different approaches have been deployed by urban scholars. My hope is that thoughtful investigations grounded in specific literatures combined with empirical work can help us to more precisely identify what exactly we intend to unlearn, provincialize and theorize anew.

Chapter 8 begins with my work on two concepts: environmental justice and urban appropriations. I argue that the concept of environmental justice and the causal relationships embedded in most uses of the term appear to be largely relevant to South African environmental scholarship. However, in practice, most published work on environmental justice in South Africa focuses on a single case that looks on the surface similar to iconic US American cases: a community of colour living next to polluting industry. As a result, the variety of cases of environmental injustice in South Africa are largely absent from the scholarly literature. I argue that we do not know how we might think about environmental justice if we expanded our scholarly gaze to this greater diversity (see Lawhon, 2013a). My work on urban appropriations similarly calls attention to the limitations that come with selecting cases that look like established theory, and then reading them through this lens. With co-authors (Lawhon *et al.*, 2018c), I argue that Lefebvre's and Bayat's articulations of urban appropriation and quiet encroachment might well capture the dynamics of some urban cases in South Africa. We also believe that there is a third version present in practice, but absent from the literature. We call this agonistically transgressive urban appropriation. In sum, this chapter argues that we might learn anew from trying to explain cases that do not look like   ·
established northern theory.

Chapter 9 works to distinguish between concerns with analytical claims (what is there and how it works) and normative concerns about what a city ought to be. I again draw on my own work in two different fields: urban infrastructure and the nexus of work, income and virtue. The modern infrastructure ideal has been identified as a widely held social and political goal of providing universal, uniform, networked infrastructure (Graham and Marvin, 2001). Many have criticized a scholarly focus on "formal" infrastructure, for this focus means that we rarely understand how most urban residents actually access urban services. Such work makes an important contribution in shifting our analytical focus towards new empirics, a line of argument that parallels that outlined in Chapter 8. With colleagues (Lawhon *et al.*, 2018a), I have urged us to build on this literature to also question the value of the normative goal of uniform, networked infrastructure. This is not to reject a goal of universal access to urban services, but to open our imaginations towards alternative modes of delivery.

I also write in this chapter about how my ongoing work on waste and work pushed me towards articulating norms about work, income and moral virtue underlying much of our urban scholarship. Here, the norms that have shaped my own work (and that of many other waste scholars) are much less evident, less explicit. In searching for a way through two unsatisfactory positions (waste work is not good work; waste work is better than no work), my searchings sprawled far beyond waste studies and into an interrogation of the "Protestant ethic," which suggests that there is virtue in working. I argue that this normative judgment underpins much of our contemporary scholarship and shapes our imaginations of alternatives. I connect my own unlearning with recent literatures on cash transfers, what Hulme *et al.* (2012) call the "development revolution from the global south." I examine how letting go of a moral belief in work opens up new imaginations and politics of distribution (see Ferguson, 2015). More broadly, this chapter seeks to demonstrate how identifying and reworking our norms can enable new understandings and possibilities for creating more just cities. I also grapple here with ethical questions about the role of the researcher in imagining and instigating change. In sum, in Chapters 8 and 9, I argue that more precisely identifying the components of critique might well help us to clarify points of agreement and dissent as well as navigate the relationship between southern empirics, northern urban theory, universalizing norms and learning anew.

My hope in writing this book is that it contributes to advancing more rigorous insights and lines of argumentation as scholars continue to work through the southern urban critique. I am deeply indebted to the scholars whose work I draw on throughout the book. While at times I differ from or seek to further develop specific arguments, I seek to do so in a spirit of constructive engagement. The ideas that we collectively are working through are difficult to think about, difficult to write about and difficult to make clear to others. I conclude *Making Urban Theory* with reflections on the importance of attending to the process of theory-making as we work through the ongoing challenges of attending to southern cities in urban studies. This conclusion is oriented towards the future, and the urban studies we might aspire to produce and participate in after the southern urban critique.

Notes

1 Given another lifetime, I would delve more deeply into the variegations of southern, post-colonial, subaltern, indigenous, feminist, black, intersectional and decolonial thought. What comparative texts there are tend to emphasize that these different bodies of literature are rooted in different intellectual lineages and empirical emphases (including the axis of identity, location and empirical object) rather than providing a deep interrogation of complementarities and contradictions. I found, for example, Broeck and Junker's (2014) style closer to juxtaposition of genealogies than helping a reader find compatibilities and points of dissent. My reading is that there is much commonality across the heart of the line of argumentation in these bodies of scholarship, and that central to each is a critique of the hegemony (not always the veracity) of (modernist, universalizing) northern thought. Wishing for (but unwilling to initiate) a more encompassing umbrella term, I adopt the less-satisfactory approach of using the terms at hand and primarily deploy the terms as used by intellectual communities examining similar empirical topics and literatures. This means typically using postcolonial as a reference for the process of examining the hegemony of western thought and southern theory as a push beyond this critique towards developing theory anew, mindful that no postcolonial scholar ever argued that critique was sufficient. "Decolonial" is increasingly being used in the scholarly community in reference to theory and scholarly practice, but I find that the more established term in the English literature, and particularly the repoliticized and rematerialized version articulated in geography (e.g., McEwan, 2003) still works equally well

for capturing this point. I note here Tuck and Yang's argument (2012) "decolonization is not a metaphor" and primarily use the term postcolonial for the scholarly mode of enquiry deployed here. I similarly use terms such as "global south," "EuroAmerica" and "the West" largely interchangeably, drawing primarily on their usage in the works under examination. I am mindful that all these labels are limited and discuss their strategic use further in Chapter 1. In this spirit, I deploy lower case letters for global south as well as racial terminology while also recognizing ongoing controversy particularly over the latter (e.g., Tharps, 2014; Clark, 2015). I adopt the lower case specifically in contrast to the capital letters used by the apartheid state and in keeping with scholarship emphasizing the construction of race (e.g., Shelby, 2007; Hall, 2017; Gilroy, 2019). I also use both "black" and "African" as adjectives, not nouns.

2 I use the term juxtaposition to distinguish the argument from ongoing conversations around comparison (e.g., Robinson, 2011; Hart, 2018). Here, I mean a simpler process of putting two things next to each other, and use other terms to build on the consequences of this proximity. More broadly, I am concerned that at times the wider literature on urban comparison does not sufficiently emphasize the importance of deeply examining the premises on which comparisons are built, and therefore see my argument as a postcolonial approach that is largely complementary to this work.

3 Feminist scholars have recently draw on public discourse to articulate and interrogate "comfort feminism" and urged feminism into less-comfortable political terrain, including deeper analyses of intersectionality (Moss and Maddrell, 2017). I prefer the term "at ease" rather than "comfort" for several reasons, although I find some utility in this wider frame. As a geographer, I find its spatial etymology attractive: Google tells me "ease" is "from Old French *aise*, based on Latin *adjacens* "lying close by." Ease captures the dual meanings of simple and safe. I also find that "being uneasy" has less-negative connotation, and more easily moves between the (false binary of) emotion/cognition. I propose unease as a state of critical skepticism rather than personal discomfort.

4 Generally speaking I think we often give insufficient attention to the impact of stress on our ability to think and write well (the slow scholarship movement is useful here, see Hartman and Darab, 2012; many works focus on the impact of slowness on the quality of life of academics rather than the quality of scholarship, e.g., Mountz, 2016). One simple measure I have seen is that people perform less well on IQ tests when stressed, and surely this is indicative of wider impacts of stress on scholarship (*NY Times*, 1983; Solman, 2018). I return to this in the conclusion.

1 Postcolonialism and urban studies

Shifting the gaze from the city to the academy

It is by now well-established that urban theory has been developed from a limited number of urban cases in the north, and that this has problematically resulted in a tendency towards universalizing theory that fails to adequately explain urban diversity and specifically cities in the global south. This is not to say there is no long history to the study of southern cities (see Myers, 2003; King, 2006), but these studies have largely been seen as sitting on the margins of urban studies. Further, scholars of southern cities participating in "international" conversations in urban studies have been required to frame their analyses through this northern-based literature (and often explained as exceptions to the northern norm), rather than develop new points of entry (Sanders, 1992; Myers, 1994; Robinson, 2006; Sheppard et al., 2013; Robinson and Roy, 2016). In contrast to earlier studies of postcolonial cities (e.g., Jacobs, 1996; Yeoh, 2001), my focus in this book is a specific intellectual move that turns the postcolonial critique back onto the academy and focuses on the conditions of the production of knowledge about cities (e.g., Robinson, 2006; Sheppard, et al., 2013; Roy, 2016). Mindful that there has been a conflation between the study of southern cities and this specific line of argument-ation[1], I use the phrase "the southern urban critique" to describe this analytical focus on the process of urban research and theory-making (Lawhon and Truelove, 2020).

Not all scholars articulating and building on what I call the southern urban critique use the term "global south." As with any category, this term has a history, multiple usages and plenty of limitations. Older works use phrases such as "Third World" or "developing world," and many of the scholars discussed deploy continental-scaled regional frames (i.e., Asian, African, Latin American cities). To provide some measure of consistency in this text, I typically use the term "south" even in discussing this longer history. As with all regional geographies, it both has useful (if partial) explanatory power and, if reified, provides a deeply flawed understanding of the construction of spaces and places. The global south is not a neutral spatial category: it and its predecessors bear the long weight of fraught construction. And yet, recognizing the limitations of any category, there is analytical utility in naming this imperfect concept (Roy, 2014). Here, then, I use this term not to denote an essentialist spatially bounded unit but to identify a dynamic, constructed concept (Dados and Connell, 2012; Wa Ngugi, 2012). Like all categories, the utility of "the south" will change; we can already see its declining salience as "middle-income" countries such as India and China push at the edges of its utility for understanding our contemporary world. For now, it is an at-hand term with its limitations; my belief is that, at present, the term retains tactical *analytical* and *political* utility in the project of critiquing and changing knowledge production. Equally, throughout the book I use the term "northern theory" mindful of its limitations and with the recognition that most scholars, even those based in and studying northern cities, are at least somewhat influenced by spatially plural ideas, people and places.

In this chapter, I seek to provide a longer history for the southern urban critique. Having recently published a piece in *Urban Studies* which simply notes a longer history because of the limits of space, here I take advantage of the luxury of a monograph to delve a little deeper into the roots of this concern. I am interested in understanding the different arguments that emerge from different authors as an entry point into the next chapter which seeks to disambiguate the southern urban critique. As described in the introduction, I also attend here to the intellectual politics of who can and ought to write about southern cities.

The emergence of the southern urban critique

Scholars of postcolonialism have long examined the production of cultural knowledge as well as the production of scholarly ideas (e.g., Said, 1978; Godlewska and Smith, 1994), including consideration of the relationship between researcher and subject (Spivak, 1999; Sidaway, 2000; Noxolo, 2009). In the early 1990s, as postcolonial theory was becoming part of the lexicon of development geography (Gilmartin and Berg, 2007; Sharp, 2009; Jazeel, 2019a), a small number of geographers sought to relate postcolonial ideas to the construction of knowledge about cities in the global south. As is true for any intellectual conversation, there was surely more interaction between various scholars than is apparent in the written published record; here I review a few publications mindful that the emerging dialogue was happening through and beyond such texts.

In 1991, in a Presidential address to the Canadian Association of Geographers, McGee drew on decades of work in Asian cities to reiterate the need to think differently about southern cities, not to transpose established perspectives onto the dynamic sites of urban change in the global south. While he adopts postcolonial vocabulary in his identification of Eurocentrism in this piece, the essence of McGee's argument long predates postcolonial theory in geography; the preface to his 1971 book – no, that is not a typo – says, "The theme of the essays in the first section is my growing disillusionment with the application of the theories that have emerged from the study of the urbanization process in the West" (McGee, 1971: 9). Similarly, Slater (1992) argues against the use of northern theories in his examination of Eurocentrism, arguing that the south is different and therefore requires new theories. He specifically thought through the relevance of Lefebvre and urban geographical ideas that build on his work. As I develop further in Chapter 2, these arguments are part of a broader narrative that positions the south as empirically different, and calls for theoretical approaches that account for this difference.

Around this same time, Sanders (1992) published a piece in the "Debates" section of *Antipode* called "Eurocentric bias in the study of African urbanization: a provocation to debate." Like McGee (1991), she points to specific urban theories and their limitations in explaining southern cities (in his case, Asian; in hers, African). For both, ideas generated from EuroAmerican urban studies provide a basis for comparison, but fail to explain the data collected in southern contexts. Sanders, in an argument that bears much resemblance to recent work (e.g. Robinson, 2010), notes that African cities are seen through northern lenses and ultimately judged inadequate/inferior or un-urban.

As best I can tell, Garth Myers was the only one to respond to this provocation in text. Myers' abstract, published when I was in eighth grade, could equally have been published last year with a few minor terminological updates: "A number of geographers have recently championed the struggle against Eurocentric theoretical categorizations in geographical research on African cities. At the same time, many whose work has concentrated on African urban geography have felt left out in the more abstract theoretical debates of their colleagues based in the West. I argue for the possibility of confronting Western bias and contributing to broader theoretical debates by creating theoretical constructs derived from the African experience" (Myers, 1994). It is worth noting that Myers' article *starts* with the premise that we all know the limitations of thinking through African cities with north-derived theory and provides citation rather than extensive examples. In other words, several decades ago this idea was not articulated by Myers as novel news. His literature review, however, looks rather different from more recent scholarship as it focuses on responding to the rarely cited Sanders invitation to debate. Specifically, Myers' takes issue with Sanders' (1992) explication of how we might undertake African urban scholarship as well as who might be positioned and permitted to do so.

In her provocation, Sanders (1992) points us to Asante's (1988) "The Afro-centric Idea" and argues for a new relationship in which Africa can be the subject, not object, of research. Sanders does not give us much detail about Asante beyond this, but Sanders' reference to Asante as a starting point through which to rethink African urban studies becomes the focal point for Myers' response. Myers pushes back against two implications that he attributes to Sanders through a more explicit articulation of Asante's Afrocentricity. The first is the privileging of "emic" perspectives (i.e., focusing on the logics internal to the city/culture examined). While Myers gives limited space to this concern, he here parallels wider literatures fearful of what Hountondji calls "ethno-philosophy" (Hountondji, 1996; see also Connell, 2007). The second point Myers contests is that those best able theorize African cities are black African scholars. Myers (1994: 198) observes, "Though he rejects the nationalistic argument that 'all teachers of Black Studies should be black,' Asante does so only to contend that these black teachers must be the right kind of blacks, i.e., Asantean Afrocentrists." Sanders herself does not quite explicitly oppose outsiders studying African cities, but there is an interesting parallel in her observation that Africans often rely on "Western interpretive models" in their scholarly analysis, but that "A perusal of recent dissertations on Africa and articles in African Studies Review reveals that many young Africans do not consistently reproduce conventional urban research" (Sander 1992: 211).

Myers instead draws on Geertz and the anthropological tradition to urge human geographers (of any race) back to the field, and to use African cities as sites from which to generate concepts rather than rely on northern theory as our intellectual foundation. While in agreement with the need for concepts with relevance beyond African or southern cities, I am sympathetic to why scholars such as Sanders might be unconvinced that spending time in the field is enough. I am, therefore, in agreement with Myers' opposition to prioritizing emic approaches as well as the question of who ought to do work on/in Africa, although I find that his recommendations in this particular piece do not take us far enough in our analysis of the process of theory-making (one can only do so much in any article!). As I have detailed in the introduction, there is utility both in concerns about who might be best positioned to theorize African cities as well as the importance of field time. And yet, I understand both as *proxies* for what I argue more substantively opens us up to thinking differently about African cities. Racial identity as well as time in the field enable, but do not directly cause (are neither necessary nor sufficient for) unlearning and learning anew. It is the slow and often fraught process of identifying assumptions within theory and within our own gaze that requires our explicit attention.

Myers was my first real introduction to African studies, and yet, somehow, I am as guilty as any in missing the longer roots of our contemporary conversations for too long. For despite thousands of citations to later work on the southern urban critique, Sanders and Myers' conversation is woefully undercited. As in, according to Google scholar, less than fifty citations between the two. Slater's work has been better cited, but it is less explicitly urban in its framing, and is cited largely in nonurban work; McGee's wider ouvre is fairly well-cited, although the Presidential address noted above which is explicit about the limitations of northern theory has also received less attention.

I have, of late, wondered why this might be. I wager that most of us trained in urban geography missed this set of scholarly interventions from this moment. McGee and Slater both are better known as development geographers than urbanists; McGee, for example, is included in a collection on key thinkers in development studies (Lea, 2006). Garth Myers tells a lovely anecdote about the surprise of his department when he identified his ability to teach an urban course: he is an Africanist, and no one really thought much beyond that. Of the authors reviewed here, Sanders in most clearly identified as an urban scholar; as best I can tell most of her work is focused in the global north. Maybe in the early-1990s urban geography and urban studies more generally were not ready to mainstream this conversation. Maybe arguing against the canon and enquiry about who can produce this knowledge was more than mainstream urbanists could quite handle. Maybe we needed a softer entry point.

A decade later, Robinson provided us with what, in my reading, is a softer entry. Her 2002 piece in the *International Journal of Urban and Regional Research* (notably, an urban journal, in contrast with the outlets for the works cited above) starts with a related concern about our limited understanding of southern cities. Here, and later in her more detailed monograph, Robinson (2006) frames her concern in terms of disciplines rather than individuals, nationalities, race or indigeneity. She explains the gap between northern and southern cities as primarily rooted in different disciplinary attention: while northern cities contribute to and are understood through *urban* theory, southern cities are largely examined through conceptual frames from *development* studies. She traces this back in part to work on dependency theory in Brazil in the 1970s, and geographer Milton Santos' urgings that cities be contextualized within this global political economic structure. This explanation largely accords with the consideration of McGee and Slater as development geographers rather than urbanists noted in the previous paragraph.

Yet, in contrast with others, Robinson's proposal is to draw southern cities (back) *into* urban studies, allowing for "ordinary cities" (such as, but not limited to, cities in the global south) to encounter and contribute to urban theory. In this move, she is equally interested in thinking more carefully about how northern theory might be usefully deployed in southern cities. In other words, her central contention is actually substantively different from Slater, McGee, Sanders and Myers: the problem is not about southern difference and too much northern theory. The problem is the designation of southern cities as part of *development* theory and not enough *urban* theory. (I examine Robinson's argument in more detail in Chapter 3).

In the burgeoning of attention to southern cities to that has followed Robinson's mid-2000s argumentation, however, I believe that key points of difference between these various scholars have regularly been conflated, even where scholars are not going back to read and cite earlier works. Slater (1992), McGee (1991), Sanders (1992) and Myers (1994) place much more weight on the difference of southern cities and the implications of this difference for how we theorize. Sanders' line of argumentation focuses on from what and whom new theory ought to emerge, but is similarly concerned with north/south difference and *too much of the northern gaze*. Robinson is as equally emphatic as these authors that urban theory can be made from southern cities, but the novelty of her central thesis is the *distinction* she makes from this longer line of enquiry: southern cities are not so different that they ought to be studied as a separate category, through separate theoretical constructs, whether African, Asian or "Third World." And, while neither the 2002 nor 2006 pieces address the point directly, we can draw on her wider work here as well as both before and after these publications to suggest that, contra Slater, Lefebvre just might be useful for thinking about southern cities. In the next chapter, I tease out these distinctions between different lines of argument in more detail through an investigation of the current literature.

Note

1 As but one example of how the literature thus far fails to make distinctions between the study of southern cities, postcolonial urban studies and the southern urban critique, I point to a recent edited collection. Edensor and Jayne's (2012) introduction to their edited volume *Urban Theory Beyond the West* describes what I have called the southern urban critique, but their literature review includes and makes no distinction between cases that uncritically deploy northern theory and those that take it as an object of critical enquiry.

2 Disambiguating the southern urban critique

With Yaffa Truelove

The southern urban critique has instigated renewed scholarly attention to southern cities and placed the south and theory-making more explicitly into urban analyses. As with any emerging set of ideas, however, there remains ambiguity, uncertainty and difference amongst advocates of this wider movement: what precisely is being argued is less clearly collective. Robinson and Roy (2016: 185) describe such "investigations into global urbanism" as "a heterodox field of inquiry which, in the last decade or so, has been tremendously enriched by lively debate, a proliferation of paradigms, and experimentation with various methodologies ... [such studies] experiment with new possibilities for a more global urban studies, to work with but also press at the limits of extant urban theorization and method and at the same time to explore the potential to start with some entirely different resources and places." Beyond this generalization, however, there is much uncertainty about the foundations of critique and their implications for scholarly practice. This chapter seeks to echo Robinson and Roy's point that the southern urban critique is not a singular call, but instead an umbrella term for several not-quite-disparate threads of argument. It also positions the focus of this book – the process of unlearning and learning anew as central to postcolonial theory-making – within this wider call.

As is often the case with writing, this chapter is rooted in my own struggles to find my way through the extant literature. During my postdoctoral fellowship at the African Centre for Cities (ACC) at the University of Cape Town, I helped start a reading group on African cities. In hindsight I better understand that I was interested in working through the southern urban critique from an African perspective, but at the time struggled to name this body of literature. My hope is that this book, and this chapter in particular, helps others to find their way through this wider conversation with more ease – and more quickly than I did!

This chapter builds on a piece published with Yaffa Truelove in *Urban Studies* (Lawhon and Truelove, 2020; in this chapter "we"' refers to Yaffa and I. And, yes, it took that long to work through ideas, although there were also two babies and a few new jobs between my time at the ACC and the publication!). We are certainly of the mind that at this moment, the heterogeneity and exploration described by Robinson and Roy (2016) is healthy for advancing the broader collective agenda and our intention is not to suggest a need for reconciliation, unification, definitive claims or ideological close-mindedness. We find utility in the agonistic processes that can result from difference and believe that this contributes to our scholarly thinking. But we are concerned that much of the engagement with and critique of southern urban thinking is based on acrimonious defences of intellectual territory rather than engagement with and on the terms of the other (Roy, 2016; Hart, 2018). Thus, we seek to clarify arguments developed through the southern urban critique including points of convergence and disagreement within this intellectual community. No doubt there are other ways to articulate and arrange what follows (for example, Pieterse, 2012 usefully similarly describes what we here call Propositions 1 and 2 as a basis for southern urbanism; we draw on his work but seek to make slightly different set of distinctions). We are mindful of the simplifications of turning complex theory into single statements. Our analysis is thus an effort in the spirit of constructive dialogues of difference.

We have found some difficulty in our review in teasing out lines of argumentation in specific works: in part, we believe this is because authors do not always distinguish polemical tactics from truth-claims, but also because authors often interweave differing lineages and strands of thinking within any given piece, making the extrication of analytical strands a challenging process. We recognize the limitations of our inferences under these circumstances. We are also mindful that, in developing new sets of ideas, author's positions may develop and change over time: we do not interrogate consistency or trace the evolution of thought over time but instead focus on delineating and disentangling specific arguments. Here, I share the four propositions identified in Lawhon and Truelove (2020) regarding the source of the southern urban critique. While some scholars' work resides clearly within one or another proposition, this paper shows how differing iterations of the southern urban critique can, and have been, productively mobilized *in tandem*.

The null proposition: speaking from the south is an argument against (all/northern) theory and in favour of particularism and empiricism

We begin with this null claim in order to insist that *we find no evidence of urban scholars articulating this position* (although Jazeel, 2019b provides a complicated reading here). We name it to clarify where the southern urban critique has been misinterpreted (e.g., Roy (2016) argues that Scott and Storper (2015) misread her own work), and to distinguish it from the propositions below. Scholars of southern cities have indeed been critiqued for an excessive focus on data, or "objectivism" (Brenner *et al.*, 2011; see also Bernt's (2016) problematic framing of Ghertner's work on gentrification as "very particular"), for not linking findings to larger (structural) explanations and frameworks. More widely, scholars such as Robinson have been misapprehended as proposing "an urban studies which is idiographic, provincial, nominalist, and comparative" (Smith, 2013: 2301).

Leaving aside the final term, we (with others, e.g., Roy, 2016) argue that this is a misreading of the southern urban critique. We do not disagree that many studies in and of southern cities, including and beyond those cited by Brenner *et al.* (2011), are largely empirical. Yet, importantly, the authors of these texts *do not argue that this empiricism is the logical endpoint for such studies;* it is our contention that, equally, they do not believe this is the logical endpoint.

Why does it continue to be the case that it is difficult to theorize southern cities? It is important to bear in mind that the vast majority of research has been and continues to be done in northern contexts (and that there is inequality within this grouping too, for we understand more about Western than Eastern Europe, more about big cities than small ones, e.g., Ferenčuhová, 2016, Bell and Jayne, 2009). This is no doubt linked to tremendous global inequalities in resources, limiting the number of scholars based in southern institutions and their ability to fund and make time for deep scholarly engagement. The weight of colonial educational systems (which had different policies, practices and local impacts over time, space, race and ethnicity) also continues to shape global knowledge production. It is also worth noting the impact of reviews and audience: imagine the limitations on writing about New York if each article had to describe its colonial history, provide a population count and familiarize the reader with the American political system. In sum, for both intellectual and institutional reasons, while we know more about some places than others, most of the global south remains particularly understudied or at least underrepresented in international academic forums (Caillods, 2016; Aplerin, undated).

This can contribute to a tendency in the literature to overgeneralize from small studies and hastily develop (inappropriate) policy recommendations (Robinson, 2002; Pieterse, 2011). Further, the politics of knowledge production has largely contributed to a situation in which substantial effort has been put into critiquing the shortcomings of existing theoretical explanations as a precursor to developing alternative explanations (see Mabin, 2014). In this context, there are very real concerns as to what constitutes sufficient data from which to responsibly and rigorously theorize (Pieterse, 2011). Important to our analysis here, *these are questions of degree, not essence:* more empirics are needed not as an endpoint, but as the basis for deep understanding, context-based analysis and more rigorous theorization (although as we explore further below, there are differences within this intellectual community regarding what precisely it is possible to theorize, again, see Jazeel, 2019b).

Finally, it bears repeating that we find no evidence of the southern urban critique rejecting northern theory in totality. Foundational, well-cited authors even use the ideas of northern scholars to develop their critique (e.g., postcolonial theorists Said and Chakrabarty draw extensively on Foucault)! Robinson works with Deleuze and Lefebvre! What underpins these concerns is not a renunciation of learning and theorizing across space, but a demand for more careful consideration of how theory travels, and what can/ought to happen when it does (Said, 1983; Clifford, 1989; McFarlane, 2012; Bunnell, 2013; Lawhon *et al.*, 2016).

Proposition 1: the south is empirically different

This proposition is based on the argument that southern cities are socially, materially, culturally, politically and/or historically different from northern cities. Here, the south is deployed as a geographical location that is relational, although any neat lines have been widely challenged. This argument is longstanding in the study of southern cities, as Chapter 1 has shown. This idea of difference is evident in efforts to define "African," "Asian" and/or "Third World Cities" as distinct categories (see Chapter 3).

Demographic arguments about increased urbanization trends or arguments about institutional inequalities across universities, for example, are largely rooted in the deployment of the south as a location, albeit one that is relationally produced. Schindler (2017: 48) and Watson (2009) are emblematic in arguing that empirical difference matters. For example, Schindler proposes a set of "tendencies" that can be seen across a number of southern cities. These include urban governance regimes geared toward the transformation of territory (rather than populations), dynamic metabolic configurations and the co-constitution of materiality and political economy that upset explanations that privilege either context or theory above the other (expressing the simultaneous need for both). These different empirics, he argues, require different theories and even a different "paradigm" to explain them: rather than deducing explanations from existing frameworks, more data are needed as the basis for inductive theorization that can then be mobilized to counter, amend and create dialogue with theories derived from the north.

Some studies that forefront southern difference do so as a means to displace concept-analytics and ideologies derived from the north. Rather than elaborating something similar to Schindler's "tendencies" that can be applied to cities of the south more broadly, other authors provide tailored explanations of phenomena in cities or sets of cities within one country. For example, Parnell and Robinson (2012) take to task the idea that neoliberalism is an overarching ideology that explains urban transformation in South Africa. The authors do not deny that neoliberalism is a political economic philosophy that may well have advocates everywhere, and that it may well influence policy everywhere. In fact, they concede that it may be reasonable and productive to consider neoliberalism and its impacts globally. However, their central contention is that even where neoliberalism is prevalent, its explanatory power is not ubiquitous, and at times the concept insufficiently explains the causality of urban outcomes. Here, approaches that place neoliberalism as the *deductive* cause prevent an *inductive* view of other drivers of change (see also Lawhon *et al.*, 2016).

In a parallel vein, Ghertner (2015) argues that different empirical conditions mean that theorizing large-scale displacement in Indian cities as gentrification obscures meaningfully different processes undergirding spatial change. He demonstrates that most violent forms of displacement are connected to enduring legacies of large-scale public land ownership, common property and mixed-tenure informality. By shifting his analytical gaze, Ghertner shows a wider and more case-specific repertoire of political possibilities for mobilizing against displacement. In line with Parnell and Robinson, he argues that these empirical differences inform not only scholarly analyses of what is there, but also the politics of possibility. Exporting the concept-analytic "gentrification" to Indian cities is not only insufficient for revealing a set of logics at work that fuel spatial transformation: exporting this theory also inappropriately guides political action (see also Lawhon *et al.*, 2014).

Empirical difference also includes analyses of what might be possible for specific cities. For example, some use historical analyses to show that it was precisely through the causal processes of colonialism and capital accumulation within imperial centres that northern cities were funded with universal infra-structures (the modern infrastructural ideal) just as colonial cities were extended fragmented services (Sharan, 2011). Thus, there is a critical attention to how the south is differently embedded in global flows (both historically and in contempor-ary times), with questions regarding whether modelling the north is feasible given the current financial, political and material terrain of southern cityscapes. There are important theoretical implications of this: such dynamics and historic rela-tions must be rewritten into northern accounts. But the motivation for articulation and the significance of the argument here is more than theoretical: from such analyses, scholars argue for the rethinking the appropriateness of particular urban models and policy goals. This does not mean conceding goals such as the provi-sion of basic services, but instead that they must be achieved through alternative modes of urbanism. New theorizations are required to understand the possib-ilities of what might usefully contribute to new political aspirations (see also Chapter 8).

While often framed in terms of north-south differences, much of the concern raised through the lens of empirical differences echoes debates found elsewhere about the ability of scholars to generalize across difference. Robinson (2016a: 4), for example, frames the methodological challenges of urban studies after the southern critique as "only a specific case of the more general problem of develop-ing concepts through particular observations across multiple settings or instances" (see also Bernt, 2016 or more broadly, Tilly, 1984). More specific to spatial stud-ies, there are long-standing debates about the utility of regional approaches. Such approaches have been critiqued, not because they do not (at times) accurately aggregate empirical phenomena (e.g., see King, 1985 for a critique of the term "colonial city."), but because such categorizations have been argued to offer less to wider theoretical scholarly agendas (Turner, 2002), even if the constructed, tem-poral and relational category is more nuanced than old-style regional geograph-ies (see a recent special issue edited by Ferenčuhová and Gentile, 2016; Myers, 2011 and Roy and Ong, 2011 for examples of regional approaches to southern cit-ies). Finally, there are implicit parallels to grounded theory (Glaser and Strauss, 1967): there is a sense that, if one encounters new data without using existing theorizations, new explanations can emerge.

At their core, critiques of northern urban theory based on empirical difference found in the literature have scholarly parallels throughout the social sciences. They can be sufficient logical grounds for developing southern or regional clusters but are subject to much of the same critiques as regional geography and grounded theory. Such work might be understood to draw on postcolonial theory to explain the primacy of northern explanations or the paucity of data and disciplinary histories that have undervalued southern sites. Our concern with emphasizing empirical difference and rooting the southern urban critique here is not that we disagree that empirical differences are real, or that one might usefully group cities as more and less empirically similar. It is that this proposition does not, in itself, challenge the ontology of any conventional approaches to urban studies. When made alone, this proposition calls for better theorization derived inductively from southern cases to better articulate and explain the southern urban condition. We find this to be a necessary but not sufficient claim and argue that the southern urban critique is strongest when concerns about empirical differences are paired with a deeper critique of the production of knowledge.

Proposition 2: the south has had different intellectual traditions

It is well established that publishing in urban studies, as in most other academic disciplines, requires engagement with existing literatures. Implicitly, for most "international" journals, this means engagement with northern-derived conceptualizations of the city that are not always relevant or applicable to the southern condition (Mabin, 2014; Schindler, 2014; Robinson, 2016b). And yet, throughout the academy, there is a growing critique of such practices and a re-engagement with both scholarly and vernacular understandings from elsewhere (much of this recently through the lens of decolonization, see Asher, 2013; Jazeel, 2017; Radcliffe, 2017). For decolonial scholars, this is more than just a spatial version of postmodern and poststructural critiques of modernity, and the voices of subalterns (rather than, say, using Foucault or Gramsci in the south) are essential to contributing new perspectives (Grosfoguel, 2011).

Historical analyses have provided trenchant critiques of the impact of racism in the development of scholarly knowledge. For the most part, this resulted in views of non-European scholarly and vernacular traditions being studied through anthropology as accounts of cultural myths or historical artefacts (Chakrabarty, 2000). These traditions are primarily studied in rural areas (and even when studied in cities, framed as "rural" phenomena, for urban residents are assumed to be "modern"). Scholars such as Chakrabarty (2000), Comaroff and Comaroff (2012) and Connell (2007) argue that there is a need to re-view these lineages not as artefacts but as bodies of thought with contemporary relevance. Intellectual traditions from elsewhere in and outside the academy are central to expanding our understanding of the world. For example, much recent EuroAmerican social science bears resemblance to non-Western thinking. However, despite concerns about the production of knowledge, none of these authors argue for a *pluralization of knowledge* as the endpoint. This is to say, the authors writing in this vein reject relativism, and believe that research on and from the south can lead to better understandings within and across sites. Intellectual traditions from elsewhere should neither exist in parallel nor be simply subsumed into existing theory. Instead, new ideas, themes, connections and approaches ought to be interwoven (displacing or changing some existing ideas) to produce new, cosmopolitan understandings.

In urban studies, we find this argument most cogently and forcefully articulated in Roy's (2009) "New Geographies of Theory." In contrast with (although neither in contradiction nor a necessary logical extension of) calls for more empirical studies, this is a call to *investigate already existing knowledges and knowledge traditions* across the globe, including attention to "how and why particular concepts are produced in particular world-areas" (Roy, 2009). Such explanations are based on more than empirical difference: different understandings of the urban emerge from history, culture, language, research networks, conferences and so on. For example, Roy argues that the strength of dependency theory in Latin America has shaped the type of Marxism deployed by urbanists: dependency-based theorizations by Latin American urbanists can be explained in part because of the politics and geography of knowledge production rather than because Latin American cities are necessarily more enrolled in relations of dependency than others. Here, the south is again largely framed as a relationally constructed location that shapes places and theorizations of them; this iteration calls for deeper understandings relationally generated in, through and from these sites.

One might consider Roy's work as a foundational citation through which scholars of non-EuroAmerican traditions can build a justification for the inclusion of diverse theoretical lineages and vernacular models of knowledge-building (although there are longer debates outside urban studies, e.g., Sidaway *et al.*, 2003; Garcia-Ramon, 2004). For example, Ernstson *et al.* (2014) expand on Roy's framework by further developing conceptual vectors from African cities and African scholarship. Others employ a "worlding" lens (see Roy and Ong's (2011) edited volume) aimed to specifically recalibrate urban theory through attention to the locatedness of urban models, policy mobilities and governmental rationalities within and across Asian cities. Karpouzoglou and Zimmer (2016) focus on vernacular ways of engaging with wastescapes: these different ways are important to recognize both for their contributions to knowledge as well as because they are differently legitimized by the state. A range of case studies also examine the vernacular ways urbanites see and experience specific southern cities, contest modes of urban development and co-construct the infrastructures of daily life, revealing situated accounts of knowing the city (Cornea *et al.*, 2017; Doshi and Ranganathan, 2017; Truelove, 2019).

We find utility in this overall proposition for advancing the southern urban critique beyond calls for consideration of empirical difference. It is useful both for explaining how urban studies got to where it is, and as one avenue through which to root and develop alternative theorizations of cities. But we argue that these two propositions alone – based on empiricism and alternative traditions – are unlikely to radically alter the status quo (which, we recognize, is not necessarily the goal of all scholars advocating the southern urban critique). This is because the modern, rational, colonial gaze continues to shape the ways in which such scholarship is viewed and evaluated – a point we now turn to.

Proposition 3: researchers need to deconstruct their assumptions with regard to southern (and all) cities

This iteration is based on what we here call "deep postcolonial theory," a strand of postcolonial theory that insists that colonial relations and rationalities remain deeply embedded in the present (and bears similarity to decolonial arguments above; see Jazeel (2017) "Mainstreaming geography's decolonial imperative"). Understandably, there is some seeming impatience amongst scholars across the disciplines (as well as in societies globally) weary of this well-established argument. And yet, scholars advocating deep postcolonial theory continue to remind us that we are all fundamentally shaped by the colonial encounter. We simply cannot expect that reasonable, well-intentioned, learned researchers will (be able to) transcend colonial rationalities by simply looking at more data and developing a culture of openness to new insights. Nor can we expect scholars to adequately evaluate and integrate other traditions of knowledge without explicit attention to the colonial gaze. Robinson (2016b: 192), for example, rather hopefully suggests, "Starting somewhere else exposes the parochial elements of these inherited perspectives … at which it is clear something new is required." However, a postcolonial lens helps to explain why the data and concepts generated from southern accounts have, surely, not thoroughly convinced all urbanists of a need for something new and shifted our intellectual curiosity southwards. As Roy (2016) begins her provocatively titled essay, "Who's Afraid of Postcolonial Theory," "Recent assertions of urban theory have dismissed the value of postcolonial critique in urban studies." Despite the advances noted in our introduction, the study of southern cities, and particularly theorizations emerging from them, remains of secondary interest in urban studies. As but one indicative anecdote, the *Urban Studies* website names Scott and Storper's (2015) assertion of a "foundational concept" for urban studies as its most viewed article over the last year (a piece that includes a critique of postcolonial theory for its particularism, although see our null proposition above), and a total of eight of the ten of the most-downloaded articles are about northern urbanism.

Postcolonial theorists have argued that scholars must interrogate the production both of their own worldview and the worldview that produces their data (Spivak, 1999; Jazeel and McFarlane, 2010). In the case of urban studies, this means scholarly practice has not accounted for the historical, modernist, EuroAmerican heritage which continues to shape the ways in which researchers encounter not only the global south, but cities in general. As a consequence, research may take place in a postcolonial city, but remain colonial in its outlook: the selection of empirics, mode of data collection, analysis and so on may still be based on a fundamentally problematic rationality. As but a simple example, many apartheid planners saw and experienced their cities as rational, efficient, segregated order. Others saw and experienced the same spaces as places of injustice, violence, exclusion. These historical experiences and lineages call attention to the potential of very different aspects and interpretations of the same empirical phenomenon: they will shape what and how phenomena are included, categorized, explained and normatively judged.

And in this we see most clearly a second use of the term "global south." It is, as Connell (2007 drawing on Slater, 2004) argues more generally, as a term and device necessary for the wider analytical project of deconstructing knowledge-power relations. The south, here, is a "concept-metaphor," a set of ideas rather than (or in addition to) location(s) (Roy, 2014 drawing on Sparke (2007); in this sense more precise geographies like south-east might work against this conceptual understanding, see Yiftachel, 2006, 2009; Watson, 2013). It can be deployed to signify the specificity of all knowledge-theory by displacing the north as universal and the south as particular/exceptional, an analytical tactic that provincializes dominant theories underwriting global urbanism (Roy and Ong, 2011; Sheppard *et al.*, 2013).

This postcolonial perspective includes deconstructing and politicizing the idea that northern cities are better and the source of southern urban aspirations. For example, critiques of particular urban models derived from the north, such as "smart cities," demonstrate how both the conception and implementation of such modelling are a mismatch in southern cities; they can actually work to further marginalize already precarious and significantly large poor populations (Datta, 2015). Here, scholars reveal ways that inter-referencing and urban modelling work to bypass the already-existing major social justice challenges facing cities of the south. As a result, it is argued not only that this approach is ill-fit, but that its outcomes ought to be rejected because they will be socially harmful. This version of the southern urban critique does not necessarily address "theory" as conventionally imagined: as a purely scholarly exercise of understanding the world. Instead, it emphasizes the ways in which norms become embedded (often against original intent) during as theory travels particularly to southern cities. I pick up this theme on the connection between theory, norms and practice again in Chapter 8.

The colonial modern rationality that is the focus of decolonial and postcolonial scholarship cannot necessarily be overcome, but this does not mean there is nothing we as scholars can do. For example, Lawhon *et al.* (2016) examine the authors' processes of "unlearning" as a way of examining the foundational premises of how we approach our research (see Spivak, 1999). In an alternate vein, Chattopadhyay's (2012) book aptly titled *Unlearning the City* examines the ways subaltern groups transform, appropriate and co-construct infrastructure in cities, placing such practices at the centre of understanding the urban, rather than the periphery. Such authors argue that more reflexivity about of our conceptual categories, our visual gaze, our language, our constructions – the very tools we use to undertake research – is necessary for urban studies. Transcending north-south binaries, the focus of unlearning must not simply be on the who or where of theory but on the assumptions embedded in it (see Introduction; Nagar, 2014; Lawhon *et al.*, 2016). In this sense, the southern urban critique is not about promoting the work of southern scholars, *tout court*, for southern scholars may also deploy modern colonial rationalist gazes, or unreflexively deploy northern-derived assumptions. Equally, this proposition requires the southern urban critique to be seen as a critique that applies to how we study and approach all cities. It is, thus, entirely compatible with, for example, critiques of global knowledge production from Eastern Europe (e.g., Ferenčuhová and Gentile, 2016).

In sum, pre-existing categories of research shape what and how we research (an argument that has also been well-made within feminist theory and methodology). Starting from the south as a location (as a site of empirical difference or alternative knowledge traditions) does not inherently transcend this; juxtaposition can but does not necessarily serve this purpose. Taking this proposition seriously requires not just more research, different social networks or the examination of different intellectual traditions. Moving beyond, but not in conflict with the demand for *inductive* and *inclusive* approaches raised above, it argues for a fundamental evaluation of a researcher's ability to recognize and work to – always imperfectly – obviate the fundamental rationality that informs the so-called north-south encounter.

In this chapter, we have argued that the strongest version of the southern urban critique requires a deeper engagement with postcolonial theory and the specific ways colonial relations and rationalities remain deeply embedded in the present. This perspective also contributes to our understanding of the continued peripherality of southern cities and insights from them in urban studies, and the difficulty of de-centering existing urban theory. More empirical and methodologically robust studies are indeed needed not as endpoints, but in order to propel further theorization that takes seriously the actually existing contexts and practices shaping southern cities, and the locatedness of all theory-making.

Certainly, this challenge raises many questions: how can we know when we have adequately unlearned? How can we recognize whether a scholar, or indeed a body of literature, has deeply and appropriately interrogated concepts? At what stage can we move beyond deconstructions and towards the articulation of new views towards research that usefully intervenes in the world? What does this mean for urban studies as a field of enquiry?

While such questions are beyond the scope of this chapter to answer, this book seeks to provide deeper insights into how we might understand the intellectual process and its outcomes. I turn to questions about the implications of this critique for urban studies in Chapters 5 and 6. More broadly, we suggest that the southern urban critique needs to continue to engage with, and bring increased attention to such questions as scholarship moves forward in new and exciting ways.

In these first two chapters, I have sought to identify different iterations of the southern urban critique. This critique has long roots and can be traced back to scholars concerned that the global south is empirically different. For some, drawing on critical development theories, this difference is rooted in extraction and global material flows. Yet this is not the only version of the southern urban critique, although it is longstanding and may remain a prevalent understanding. Robinson's urgings, which were pivotal to the recent attention of urban scholars to southern cities, are specifically critical of viewing southern cities primarily through the lens of development, critical or otherwise. In this chapter, therefore, Yaffa Truelove and I have argued that we might understand the southern urban critique as having three different roots. One foundational proposition is that southern cities are different, although it is clear that not all scholars believe this is an intellectually productive grounding. We believe that there is truth to a broad claim of empirical difference but we are less convinced that this proposition adequately captures key concerns with ongoing scholarship. A second proposition is that there are different intellectual traditions in the global south which require attention and, plausibly, integration or displacement of other ideas in the wider field. We believe there is utility in looking beyond the northern canon, although cautious of readings that might suggest the sorts of essentialism critiqued in the Introduction. Finally, we advocate for a deeper engagement with our third proposition, that we interrogate the assumptions underpinning all urban theory. We argue that adequately considering southern urban differences requires thinking about the assumptions that have led to understandings of the south as insufficient, inadequately urban. We argue that engaging with other bodies of knowledge on equal footing equally requires us to think more carefully through the assumptions that underpin our judgments of their utility and veracity. In other words, approaching the southern urban critique through a postcolonial lens enables us to think more carefully about different empirics and knowledge traditions as we work to better understand and develop urban theory.

3   Provincializing the urban question

Against the rural-in-the-urban

How better to enquire into the making of urban theory than to investigate the roots of the urban question itself? While this scholarly question has received much attention in urban studies, my own interest in the urban question comes from a more mundane set of experiences. I cannot tell you at what point I came to understand that when many South Africans talk about "the city" they mean "formerly white areas" (I discuss the particulars of South Africa's urban apartheid further below). In Pietermaritzburg/Msunduzi, where I was a student and lecturer for several years, my peers (and, adopting the vernacular, I) would say we were "going to town" when we went downtown to the central business district. At first I internalized this as shorthand rather than a categorical distinction that the university and nearby residential areas were not "town." The vocabulary that drew my attention in Cape Town was different: I vividly remember being genuinely unclear, at first, when an older well-to-do do-gooder at a volunteer project in Khayelitsha said "like they do in the city." It soon became obvious that she meant not only "not like here in the township" but also the corollary "the township is not the city."

While these served as interesting provocations about vocabulary and geography, it was in teaching urban geography at the University of Pretoria that I realized this question needed more direct engagement and analysis (and also inspired Chapter 5 on teaching a more spatially and intellectually plural urban geography). The most straightforward reason was that I could not make sense of the students' papers without a deeper understanding of how they were using terms such as "urban" and "rural." For example, many but not all described the townships as "rural areas" while others distinguished formerly white residential areas from "the city." The challenge was not just about understanding "the" South African use of terms (most of my students were South African, although there were also a few international students); that would have made my interpretation easier. It was that South Africans varied in their use of these terms; sometimes one student's use would vary even within the same paper. So, my students and I had an open-ended conversation about the words, how students used them and how that differed from what they read in their Pacione (2009) textbook. While this conversation focused on the relationship between the vernacular and the textbook, it sparked my interest in where the definitions we encountered in the textbook came from as well. This chapter and the next focus on developing a deeper understanding of the relationships between the urban question, African cities and vernacular vocabulary about cities.

The urban question

The urban question – "what is the urban?" – is as old as urban studies. Agglomeration and densification are enduring characteristics of human settlement, but the markers and thresholds of what "counts" as the urban or the city for scholarly analysis are more than a function of density and have evolved over time. Castells and Sheridan (1977) see the urban primarily through the lens of Marxian analysis, characterizing it as a kind of machine for the reproduction of labour (and of course as a site for labouring). In parallel, urban geographers such as Smith (1979) and Harvey (1978) see the modern city as a crucial machinery for the investment and accumulation of capital, often at the cost of most who live within it. While pre- "critical turn" geographers defined cities primarily in economic terms (e.g., Christaller and Baskin, 1933; Losch, 1954), Lefebvre *et al.* (1996) and those critical urbanists who follow (e.g., Brenner, 2000; Merrifield, 2014) generally see the city as a crucial site of political dispossession as well. Indeed, much of the more recent work in urban geography that does not draw centrally on Marx still focuses on the city's political functionality and various modes of urban governance (e.g., Blomley, 2008; Derickson, 2016; Newman and Goetz, 2016). More recently, debate about the relevance of what some have called "planetary urbanism" has emphasized urbanization as a process over the city as site (Angelo and Wachsmuth, 2015). This move asserts the importance of cities and urban processes, but it does little to clarify subjects of urban enquiry, for if every locale is somehow bound up in urbanization, then the urban may have lost much of its analytically distinctive heft (Krause, 2013; Jazeel, 2017).

While many urbanists have disagreed over the answers to the urban question, others have suggested that there may well never be a satisfactory answer. Sanders (1992: 204) quotes Abu-Lughod's argument, "There is no such thing as 'the city' which conforms to a singular set of characteristics; there is no such thing as 'urbanism as a way of life' which leads to predictable forms of social interaction; and finally, there is no congruence between the city as a physical place and the social life that takes places within it." It may be of interest to note that Abu-Lughod of course is one of the few scholars long recognized as an *urbanist* (i.e., rather than primarily a scholar of development, see Chapter 1) who has long engaged with southern cities, and so this provocation bears particular weight here (see also Chattopadhyay, 2012 and Cinar and Bender, 2007 whose introductions also urge away from a universal definition, and also draw on cities across the north and south). As with assertions of planetarity, however, a null claim offers us little help in developing urban studies and urban theory. Despite such ambiguity, the term "urban" continues to be used largely without question.

Undoubtedly, one could usefully fill pages with a history of scholarly approaches to the urban question, and yet, in order to move on to the task at hand, this is but a brief account. Of importance here is the point that, for the most part, despite much criticism, there is and remains an implicit working definition in and beyond urban studies of the urban as that-which-has-already-been-deemed-urban. This implicit definition seems to adequately satisfy the demands of peer review and scholarly regard. And yet, I write this chapter because my own experiences in and outside the academy suggest that the further one is from the Euro American global north and the sites from which most urban theory was developed, the more troubling these conflations become in scholarly analysis (Pierce and Lawhon, 2016).

While scholarship focused on southern cities has in some cases interrogated specific urban theories (Parnell and Robinson, 2012), shown broadly the northern bias of foundational urban scholars (Slater, 1992; Robinson, 2006) and articulated other scholarly traditions (Roy, 2009), the question of what counts as "urban" has largely been elided (see below on this question in Robinson, 2006). The argument I develop below is based on the premise that it is not just that the cities from which we see are different, but that we do not all know and see the same socio-spatial composites as "city." This chapter and the one that follows attempt to see what happens if we think about the urban question from unexpected places and of unexpected persons (methodologically paralleling Myers, 2011).

In this chapter, I first seek to show the relationship between African cities and the urban question-as-a-question. In short, in the scholarly literature, there is not much of a relationship. Instead, as in much of the wider urban studies literature, the urban question is treated a something already satisfactorily answered. In contrast to much urban studies literature, *definitions of urbanization* taken from urban studies have long been central to African urban enquiry. In other words, a key point of enquiry in the study of African cities has long been whether particular sites can be appropriately deemed "urban." Here, I provide a more detailed historicization, but agree with Sanders' (1992: 201) summary: "These [investigations] have turned out to be little more than exercises in speculation, epistemological forays which serve more to reify European spatial and cultural forms than to advance thinking about Africa." As but one example, Robinson (2006: 91) raises the question of what a city is in a brief examination of Koolhaus' work in Lagos. After telling us that he considers Lagos not quite urban, Koolhaus explains that he uses a conventional urban analytic because it is the closest concept to the empirical context that is at hand. African cities, thus, may be places where urban theory fails, but such conclusions are not used to challenge the essence of what is (not) city. In other words, the conventional conclusion is that there is something wrong with African "cities," not something wrong with urban theory.

This enquiry is important because the rather one-sided way in which definitions and characterizations of the city have flowed from north to south continues, even in scholarship on the southern urban critique. While scholars have sought to provincialize specific urban theories and bring southern cities into urban studies, the concept of "the urban" has largely remained intact. In the second section, I go back to the urban anthropological work of the mid-twentieth century as it is a key literature that Robinson (2006) points us to for thinking differently about more cosmopolitan urban studies. While Robinson is clearly sensitive to positioning African cities as "failed" or "not quite urban" (see also Robinson, 2010), I seek to advance a conversation prompted by her work through an excavation that focuses specifically on the use of definitions of the urban in African literature. I seek to show the ways in which the concept of "the urban" is interwoven with a colonial modern gaze and colonial modern politics both in this older literature and at present. In contrast with Robinson's (2006) efforts to rehabilitate the meaning of modernity (so that African/ordinary cities can also be recognized as cities), I am concerned that modernity continues to be integral to urban definitions as deployed in the south, even in urban studies that have explicitly rejected such conflations.

This chapter and the one that follows work together to make the argument that universalizing northern referents continue to shape the bounding of urban ana-lyses globally (what is and is not included in urban analysis). This tendency has largely gone unobserved but continues to have significant impacts on the pro-cess of urban theory making, as well as urban politics, practices and process (Pierce and Lawhon, 2016). While these chapters focus primarily on South Africa, I believe they serve as evidence of a greater need for urbanists to explicitly attend to the way in which a universalized idea of city is enrolled in particular analyses and its impact on scholarly analyses.

The application of definitions of "the urban" to African cities

African cities have been especially under-attended as reference points within the wider field of urban studies, but the lacuna is particularly significant when it comes to the literature on the urban question. Instead of postulating what a city might be based on what one sees or says is urban in Africa, accounts typically explicitly define cities through canonical references, and then use these to delin-eate what is (not) city in Africa. O'Connor (1983) addresses this wider trend in his foundational study *The African City:* he acknowledges that the urban ques-tion has not a single answer, then notes several earlier Africanist works as having drawn their definitions from (now largely outdated) northern urban studies. While the phrase "urban question" can be found in the South African urban literature, it has been used instead as a heading for more specific questions in South African urban studies (e.g., Dewar, 1995; Visser and Rogerson, 2014; see also Roy, 2009).

Freund's (2007) historical enquiry, for example, works through the question of what counts as a city (and when a settlement becomes "city"). The criteria deployed in Freund's monograph are drawn from the accepted characterizations in urban studies developed from northern cities, such as density and economic function. Historical African cities are often described as limited and not-quite-urban; as elaborated on further below, settlements are noted to contain urban and nonurban parts. This does not make them unimportant; Freund (2007: 2) argues,

"If [sites] fail to meet certain contemporary criteria of what a city should be like, such settlements should not be dismissed but rather embraced with interest for their unique configurations and contribution to the cultural development of mankind." But they are still not-quite-urban. Importantly here is not whether Freund's descriptions accurately represent the historical data; the key for my objective here is that *an established notion of what does and does not count as city is applied, not contested or developed.* In other words, the city is the thing developed from (universal/northern) theories, and African cities are measured against rather than used to reconfigure this. The frame of enquiry is about explaining what makes Africa different and its cities not-quite urban; the frame is not a challenge to the universalizing category developed from non-African cases. African cities are not permitted to be sites from which to revise or theorize new historical and contemporary responses to the urban question.

In many other contemporary studies, as in the global north urban literature, clear operational definitions for the urban and the city are absent. Some, as in older literatures, focus on whether there is a specifically *African* city, a framing that suggests there is an accepted thing-called-city of which there are different types, such as African ones. Such enquiries acknowledge, for example, the particularly mobile lives of African urban residents (O'Connor, 1983; Potts, 2010; Myers, 2011), and the ways in which this mobility shapes urbanization patterns both in terms of spatial movement and cultural change. For example, Myers (2011, his monograph *African Cities*), and Epprecht (2016, in a more focused study of a township in South Africa *Welcome to Greater Edendale*), both ask about the *African* character of the city without defining the category of city.

The question of urban vocabulary is particularly provocative in Epprecht's work, as I reflect on further in a review of the book (Lawhon, 2018b). Epprecht is an historian, not explicitly interested in urban theory, and a person whose work and ethical engagement with the research process I deeply respect. I point here to the use of urban vocabulary in his text because it highlights a wider, troubling trend. I observe, "Specifically, Edendale is a city in the title but nowhere else: references throughout to the city are actually references to Pietermaritzburg (either the geographical area or its local government). Edendale is subject to urban renewal, but otherwise remains positioned as the 'other' to which the city is central (even the early *amakholwa* settlement of Georgetown is a village, a term generally used for black Africans, although Georgetown is noted to have been laid out as a colonial-style grid). This usage is common in South African vernacular, but retains problematic distinctions of the urban as modern, white and formal" (Lawhon, 2018b: 213). I point to this as Epprecht's use is not atypical; many scholars of African cities shift between "scholarly" and "vernacular" meanings of urban vocabulary, even within a single text.

Undoubtedly, these works on African cities challenge the idea that all cities look alike and operate in the same way and trouble overgeneralizations in much of urban studies. But they continue to implicitly start with assumptions about what city is (not), whether from urban studies or the local urban vernacular. Despite finding differences in specific aspects, these works leave intact the meaning of city as defined through various canonical approaches to urban studies primarily derived from northern referents.

Robinson (2006), in her foundational text in the southern urban critique, argues that scholars of African urbanism have at times challenged urban theory's universalizing tendencies. She specifically mobilizes the work of the mid-twentieth century urban anthropologists and their challenge of the blasé attitude (see also Robinson, 2002). While the Zambian Copperbelt studies are the most well-known, she also references investigations in South Africa that sought to distinguish between, in the terminology of the time, residents of townships who became "Townsmen" and those who remained "Tribesmen" (Mayer, 1961). Townsmen "adopt urban values" (although most of the respondents in the study reportedly have negative views of the township as crowded and dirty) while "tribesmen" exhibit few changes after their migration to the township. Robinson draws our attention to the work of Mayer and his contemporaries writing in and of African cities for their role in challenging universalizing notions of urban attitudes. And indeed, others have placed them as part of a broader movement that shifted academic notions of urbanization as a social process (Hannerz, 1980; Ferguson, 1999; Bank, 2011).

My intent in reading these scholars here is somewhat different than Robinson's in focusing specifically on the meaning of "the urban," and so, I return to the arguments present in the text itself. Mayer's (1961) assertion is actually that residents of cities who do not adopt urban attitudes (a predefined category based on research in northern contexts) are *not urbanized*. Although challenging universalizing assumptions that all those who live in cities undergo a linear and predictable realignment of practices and values, Mayer did not call into question the notion of *what it meant to be urban* in his time. Rather than challenging the northern-defined category of what it means to be an urban resident, this argument placed "tribesmen" living in South African townships outside the category. This analysis contrasts significantly with the argument in contemporaneous work in New York City. Jane Jacobs' *The Death and Life of Great American Cities* (1961) similarly advocates for an understanding of the urban as a community context of powerful relationships rather than a Chicago School-style context for Gesellschaft's weak ties. Her work, however, specifically pushes a reimagination of the urban, for neither she nor her readers doubt whether New York City is properly urban!

A similar point comes from urban anthropological investigations in West Africa around the same time. Little (1975) provides an overview of recent work, including his own writing. Here, and in conversation with the southern African urban anthropologists noted above, he is skeptical about the relevance of the established ideas of urbanization for many West African settlements. One interesting example focuses on the social structures of Yoruba settlements, which include tightly linked "urban" and "rural" components, challenging this dichotomy (this observation would be a fascinating starting point through which to challenge urban theory, and I hope this note might inspire further investigation from someone more familiar with the region!). Yet as above, the study of West African cities is not used to challenge scholarly orthodoxy. Little (1975: 15) instead argues, "We have therefore in 'urban' terms a highly complex situation which has taxed the ingenuity of numerous scholars. However, the salient consideration is fairly obvious and it is that in this contemporary African context conventional indices and distinctions provide an inadequate basis for empirical purposes. For example, Wirth's well-known minimal definition of the city is, 'a relatively large, dense and permanent settlement.' This, however, is irrelevant to the actual social reality." And yet he too lets us down here in the next statement: "The use of 'urbanization' as a concept for empirical purposes should be confined to nucleated settlements of a specific kind." In other words, some African settlements look like northern ones, and therefore count; others do not look like northern ones, and therefore, do not.

This is not a historical or spurious concern. Bank (2011), following on Mayer's work decades later, similarly describes what he sees as the rural *in* the urban. The phrase is also deployed by Mamdani (1996), although his work pushes back against the teleology implicit in typical narratives of urbanization. In other words, scholars largely assume the definition of the urban as articulated by northern the-orists and northern cases and determine that some black African urban residents act in ways that resist or defy the "real" definition of urbanism[1].

My point here is not to discredit the essence of works that use this frame, for the anthropologists of African cities sought to understand social processes; the city itself was not the object of their analysis (see Hannerz, 1980). And yet, the social process of urbanization was central to their analysis, and their framings continue to be used beyond anthropology. As such, they provide a rather fraught example from which to speak back to urban theory (as Robinson urges us to do): in challenging the teleological assertion that all people who move to cities develop a blasé attitude, they did not actually undercut the association between this attitude and "real" urban processes. To reiterate, the argument present in the texts them-selves, and often similarly deployed in the present, is that many black African migrants and settlements are different and not urban. I use these authors, then, not as a model but as provocation through which to examine what we carry with us when we create urban theory, and why postcolonial critique is so important.

The city as modern, good, desirable, white and European

The urban anthropologists reviewed above largely deployed northern scholarly ideas of the city, but it is impossible to separate these ideas from the colonial modern context in which they were developed (see Mignolo and Walsh, 2018). Specifically, the colonial encounter resulted in a conflation between the city, what was European and what was modern (Hannerz, 1980; Cooper, 1983; Mitchell, 1987; see Jones, 2000 in South Africa). As but one example, I point to the well-known work of Comaroff and Comaroff (1987) which plays with this messy association (see also Kepe and Ntsebeza, 2011 which challenges the association between urban and modern in this dichotomy). They suggest of residents of a densely populated area: "Montshiwa Township might have been nearby, but, by a rhetorical leap of the imagination, it had been placed over the border. Those who lived in it might reside in the rural area, but they were unmistakably citizens of the realm of *sekgoa* [European ways]" (see Chapter 4 for an explanation of the homelands policy and the meaning of "over the border"). What is conveyed here is that it is surprising – it takes much imagination – to think that township residents with European ways could be called "rural." Although scholars sought to differentiate between terms such as urban, rural, European and modern by providing definitions, highlighting their ambivalent scholarly usages (e.g., Mitchell, 1954; Little, 1975), in practice they often remained analytically entangled. Such conflations, importantly, were not confined to Africa but took on a particular significance through the colonial encounter on the continent; it would no doubt be of interest to also explore these questions in other literatures[2].

Controversies around Mayer's (1961) work at the time of publication usefully draw attention to the associations between the city, modernity, race, Europeanness and desirability as well as the political implications of such conflations. The antagonism I review here centred not around whether township residents really were or were not living in "the city" or were or were not "urban." Instead, Mayer's reporting of black African urban residents who retained traditional social norms was critiqued for its *political* implications: it was suggested that his findings might be used to bolster apartheid notions of the fundamental rurality of black African people (and, thus, justify exclusion from urban areas). Embedded in the liberal ideology of the time was the notion that black African people could and should modernize, and that becoming urban (both moving to a city and adopting "urban" values) was part of this project. Believing the antithesis of this – that black African urban residents did not always urbanize/modernize upon arrival to the city – was understood as reinforcing the othering promoted by the apartheid state (Bank, 2011). For example, a contemporary review by a well-respected liberal peer (Powdermaker, 1963: 474) notes that Mayer's text is important for "giving detailed data on the cultural background of a significant and often neglected problem – resistance to change – in one African society." Powdermaker explicitly identifies *resistance* to the adoption of new (urban/modern/European) norms as *a problem*. Although this represents a common reading, we agree with Bank (2011) that it is unfair to read Mayer's work as reinforcing apartheid ideas about the fundamental rurality of black Africans. I call attention to Powdermaker's review here instead because it highlights the normative entanglement of race, modernity and the urban.

In contrast with Mayer (and in accord with Powdermaker's review), the authors of the Copperbelt studies also reviewed by Robinson (2002, 2006) were largely advocates of a wider political agenda of what might best be considered in today's vernacular as assimilation (Ferguson, 1999). Ferguson (1999) provides a particularly insightful account that is sensitive to the historical context, placing the Copperbelt anthropologists as both antagonistic to the colonial project and complicit with it in their advocation of Europeanization. The Copperbelt studies differed from Mayer's work in their focus on the "full" urbanization of black

African migrants, a process celebrated by the researchers. As but one example, the Copperbelt scholars showed evidence of black Africans staying permanently in cities (as opposed to moving back and forth between the city and rural villages) and retiring in cites (instead of moving to or back to a rural village). This focus was, surely, understood to be a political tactic, for their assertion was clearly antagonistic to colonial notions that Africans are and ought to remain fundamentally tribal and rural[3]. Readings such as Ferguson's call for us to be cautious about the celebration of historical accounts that challenged urban theories, and the difficulty of articulating urbanization as a process in colonial Africa separate from the adoption of modern and/or European social norms.

It is important to acknowledge that the associations between Europeans, modernity and the city were not solely made by white colonists and researchers. These associations were and continue to be at least partially developed, recognized and/or uptaken by African people. Ferguson (1999), for example, tells us that respondents in his Zambian case described much of their worldview through what he calls the lens of an out-of-date textbook. In other words, he suggests that respondents regularly deployed concepts articulated by mid-century anthropologists that conflated race, modernity and the city in their explanation of their own context. For example, a site is explained to no longer be urban because its infrastructure is failing, it is no longer modern and its white residents have left. As with Robinson and the anthropologists reviewed above, this question was not central to Ferguson's work; I therefore seek to take on these questions more directly, building on this observation about the use of urban vocabulary in the vernacular and its relationship to academic definitions in the chapter that follows.

What happened to scholarly investigations of associations between modernity, race, Europeanization and the city in the decades since these urban anthropologies? Robinson (2006) suggests that the global studies of cities since this time has been largely defined by a separation into urban studies for the north and development studies for the south. Urban studies as a broader field of academic enquiry never conflated race, nationality, modernity and the city to quite such an extent as what is found in studies of colonial and postcolonial cities in the global south, and what associations existed have largely explicitly been rejected in theory (e.g., Dear, 2006) and as a normative conceptual goal (although aspects of modernism remain embedded in more practice-oriented areas).

And yet, in 2006, Robinson pulled modernity back explicitly into the urban scholarly conversation by arguing for a rehabilitation of modernity. For scholars based in a northern urban conversation, such an argument might well have appeared outdated or out of tune given the postmodern rejection of modernism (exemplified by Latour's claim "We Have Never Been Modern"). In contrast, Robinson's work may be read as an assertion that "We are Modern Too!" a frame that interestingly has much in common with long-standing approaches to alternative or multiple modernities in development studies (Goankar, 2001; Hart, 2002). For urbanists embedded in southern urban political realities and development studies, however, this was an impactful call because the notion of modernity continues to have widespread significance in African cities (e.g., Ballard, 2004; Popke and Ballard, 2004; Kamete, 2013; Gillespie, 2016).

How might we better understanding such aspirations to be modern, too? My sense in working through this literature and the data in Chapter 4 is that there remains analytical utility in a definition of modernity as specific set of ideas, aspirations and understandings of the world. I am mindful of the political utility of broadening this term to include multiples and hybrids. And yet, as with the term city, I am concerned that stretching it too far means that it loses its analytical utility. We need a word to describe the colonial modern outlook that was and continues to be globally pervasive. We need to understand the specificity embedded in the desire to be "modern," to have "expectations of modernity" (Ferguson, 1999), to desire the "modern infrastructure ideal" (see Chapter 9). I am excited to bend this term, to stretch it to include new spaces. I am disinclined to stretch it too far. I want to limit it so as to discard its urges towards clean categories, orderly plans, beliefs in the ability to control people and nature as well as the hierarchies of who belongs in the city. I return to the wider concern with the conflations between modernity as a phenomenon (a belief system that shapes practices), as a scholarly analytic and as a normative prescription (that we might or might not be well-served in rehabilitating or rejecting) in Chapter 9.

In this chapter, I have demonstrated that there is a long-standing tradition of scholarship about African cities and the urban question, but that scholars have largely deployed rather than developed new insights into what it means to be urban when studying African cities. And yet, I wonder – without being able to answer here – what might have happened if the scholarly community of the mid-twentieth century had been more open to considering African cities as sites from which to theorize what a city is and might be. Instead, we as scholars must grapple with the consequences of the past and a field in which the city largely continues to be, in scholarly definitions, something that remains distant from the everyday lives of many African people.

In the next chapter, I discuss the relevance of urban vocabulary to the present both in and outside the academy. I build on Ferguson's observations and the idea that the city continues to be conflated with racial and modernist imaginaries in everyday understandings of the city in Africa, and work to think through how this ongoing understanding of the city shapes urban scholarship and practice. Like Robinson (2006), my effort here is meant to push us to think differently about what counts as city or not, and for us to be open to a greater diversity of ways of being in and of the city. I believe we can look to and learn from the past, but I am equally mindful of the limitations of doing so, for the powerful presence of colonial and apartheid ideologies were and remain difficult to transcend. We cannot know how different our cities might be today if scholars and urban planners had been more open to thinking about diverse ways of being in and of the city, but we can continue being attentive to how this history continues to shape our present. My hope in these two chapters is to build on a longer lineage of scholarship to continue pushing for attention and openness to this diversity in and of the city.

Notes

1 In contrast, Myers' (In press) chapter on Chinese ideas of "villages in the city" provides a provocative point of departure. After a discussion of villages in the global south, Myers' uses of the term "village" (rather than the more conventional term "town") to describe Wetherford, Connecticut. Such an analysis is, surely, unsettling for many readers. This points us towards how comparison might, for example, enable new ideas and expose racializations embedded in much of our geographical vocabulary.

2 There is an inverse provocation in the US American vernacular in which urban can connote African-American, making the association between black Africans people and ruralness counterintuitive for many US Americans! There is surely interesting work to be done across these contexts, and this point also serves as a reminder of how race is constructed differently across space.

3 Although this position was not universally held by all actors participating in the colonial project, see Cooper's (1996) longer history of the decolonization of Africa.

4 The struggle for the city, and for the right to be city

With Anesu Makina

> We are socialized to believe that modern and well-built spaces are cities, yet who is to determine what a city is?
>
> *Interviewee, Hatfield, Tshwane, South Africa*

> I don't think Soweto is a city. It would be a huge compliment to call Soweto a city.
>
> *Personal communication, Soweto resident, Tshwane, South Africa*

In Chapter 3, I argued that scholarly responses to the urban question have largely been based on northern referents. In this chapter, I further develop this line of thinking by investigating the relationship between the cities we study, our ideas about the city and contemporary everyday urban vocabulary. While the urban question is both central and foundational for urban studies, there has been very limited attention to how urban vocabulary is used in everyday language in either the global north or south. And yet, the earliest answers to the urban question took "the city" as an identifiable site and sought to describe its characteristics in an inductive way. Subsequent approaches to the urban question, as briefly reviewed in Chapter 3, have largely sought to identify the city as an object (or process), allegedly separate from its meaning for individuals (Sayer, 1984; see also Williams, 1973).

And yet, as Sayer (1984: 279) convincingly argues, "How cities are built and how they function depends on actions and social relations which are informed by the expressive order of ideas." (We might translate Sayer into today's scholarly discourse by saying that sociomaterial urban forms cannot be separated from urban imaginaries, including imaginaries of what the urban is.) Despite Sayer's now-several-decades-old observation, urban scholars have largely elided the distinction between the "material" city and ideas of the city. Nor have they explicitly attended to the latter, likely in part because of assumptions of general accord between the two in northern cities and scholarship noted in Chapter 3.

Work on urban imaginaries has, no doubt, made important contributions (see Bridge and Watson, 2003), but this literature typically examines imaginaries within the predefined urban rather than specific enquiries into what is (not) urban. Yet it is the idea of what is urban that shapes the bounding of the subsequent analysis (what is and is not included in urban analysis), a tendency that has largely gone unobserved but has significant impacts for urban studies. Cinar and Bender (2007: xiii) are atypical in framing their edited volume as explicitly pushing back against definitions of the urban, urging the reader to suspend an effort at totality and closure. Instead, their proposal is to enquire "how cities are imagined in more open ways, more accommodating to translocal connections, internal tensions and generally loose ends, yet still recognized as distinct from the larger environs, social and physical. To locate, identify and understand such a city draws the scholar ... towards the imagination and toward the making and remaking of public culture and a. ... Shared, not unitary, mental image of the city." The works in this volume make important contributions to excavating a more plural and lived idea of the city.

This chapter takes a different approach to examining urban imaginaries through an investigation of urban vernacular vocabularies. As noted in Chapter 3, it is inspired by observations in the classroom and the African studies literature on the relationship between ideas of modernity, race and the city. I develop the argument that, if everyday meanings of urban vocabulary shape and are shaped by ideas of the city, and ideas are conjoined with the production of cities, they thus ought to be integrated into urban analysis. In line with the rest of the book, I find this argument to be particularly provocative when considered in and through southern cities. For, as noted in Chapter 3, the further we are from the cities where urban definitions were developed, the less useful these definitions appear to be for describing the "cities" or "not cities" we see. While undoubtedly scholars attempt to disentangle their everyday vocabularies and understandings from analytical approaches, it is quite difficult to see (northern) analytical biases particularly if one only uses northern referents. I believe this is particularly challenging in a place like South Africa where vernacular and scholarly vocabularies sit uncomfortably next to each other; here again a juxtaposition is useful for highlighting the limitations of each vocabulary.

In this chapter, I explore the vernacular meaning of the term "city" in one South African "urban" area as an example of the use of urban vocabulary in a southern site, mindful that it does not represent the whole south or all of South Africa. First, I provide an overview of South African urban history, including the politics and policies that shaped what was legally determined to be or not be urban. This section concludes with a review of a few terms taken from the South African census as it provides useful definitions of some vocabulary relevant to urban studies. I then provide the results of roughly two hundred short photo-elicitation interviews in South Africa about the term "city." There are, surely, many other methods one might use to deepen these insights, and my hope is that this initial enquiry might spark further work. My intention here is not to prioritize the vernacular as somehow more authentic or true, nor to argue against academic analysis; analysis of the data through an understanding of the history of cities and the ways in which scholars have studied and articulated them is essential to the wider argument.

I then argue that the meaning of "city" in South Africa in the vernacular remains deeply associated with historical colonial-modern-apartheid urban imaginaries.

Some research participants actively challenge this notion, and raise important questions about whether and how we might reconceptualize the urban in a more expansive way. I urge a deeper understanding and incorporation of this imaginary into the scholarly literature as well as politics and policy in South Africa, and further investigation into its impacts. Again, while the work focuses on South Africa and the specifics of apartheid shape the results, undertaking a similar line of enquiry elsewhere would likely demonstrate equally interesting insights.

A brief history of "the urban" in South Africa

What is a city in South Africa? Narratives about the apartheid state present a discontinuous, incongruent picture of the actual practices of urbanization across the country, but questions of who belongs to the city, and what (certain categories of) persons are (not) allowed to do in the city was undoubtedly central to apartheid imaginations, plans and practices. State narratives changed over time and were never fully adopted in all places. These shifts were in part due to recognition of the practical limitations of policy, changing economic needs and pressures, incongruence between national policy and local implementers and political compromises (Mabin, 1992; McCarthy, 1992). Nonetheless, I point here towards several significant trends as background for my enquiry into urban vocabularies as these surely have shaped imaginaries of the urban in South Africa. The language, terminology and descriptions of agency here are fraught: I try to explain the intent of policies, and to not overuse quotation marks, mindful that the text is intended to work against many of the labels necessary to explain the historical context. I also focus on black and white spaces, mindful that these were not only racial categories deployed by the apartheid government.

The residential exclusion of black African people from areas that were deemed white cities has a long history. While piecemeal segregation was common in practice prior to its enactment, the 1923 Natives (Urban Areas) Act officially and at a national scale deemed black African people to be only temporary residents of cities (Maylam, 1995), a practice in keeping with wider colonial trends (Cooper, 1996; Ferguson, 1999). The Act was followed by a series of national pass laws that required black African people to have special permits to both enter and reside in urban areas. Requirements varied spatially and temporally, and age and gender shaped their application; even so-called "permanent" passes given to those who met stringent criteria could be retracted, and individual officials used their own judgment to interpret the law. Enforcement, too, varied with space and time as well as the values and whims of individual state officials. Freehold property did exist before and after the 1923 Act, but new sales were limited. Nonetheless, as the infamous forced removals that continued into the 1960s illustrate, the apartheid vision was always fragmented in practice (Maylam, 1990 2001; Mabin, 1992; McCarthy, 1992; Smith, 1992). Further, as Mayer (1961) and Epprecht (2016) show in quite different contexts, while freehold rights were limited and much township land belonged to the municipality, privately owned houses remained, including some purchased from the municipality. It is notable that Mayer, sixty years ago, reported that lack of home ownership contributed to feelings of alienation from the city.

During the 1950s, the apartheid government developed various policies for the construction of "homelands" (sometimes called "Bantustans"), or separate "countries" for black African citizens. These newly constituted quasi-countries were intended to become the rightful long-term home for South Africa's black African peoples. Unsurprisingly, the implementation and enforcement of "homeland" policies was particularly fraught and limited but had impacts that remain today (Ramutsindela, 2007). As observed in Chapter 3, Bantustans were assumed to be rural and traditional: Jones (1999) provides a more detailed account of the conflation of urban/modern and traditional/rural in the homelands. Here I observe a certain irony that the most globally well-known site in a "homeland," then and now, is a tourist destination that bears the powerful moniker "Sun City" (see Rogerson, 1990). Sun City is largely comprised of four hotels, several golf courses and the census count of 1,299 residents.

State regulation that deemed black persons outsiders in the city did not, however, mean that black South Africans were not present in the city: indeed, the state and private actors regularly sought to recruit urban labour, to draw it to the city (Murray, 1987; Cooper, 1996). I quote Mayer's (1961: 44) narration of East London for its detail on presences and absences (and retain his capitalization style for ease of reading and as a reminder of the apartheid era convention):

> It is true that many Black faces can be seen within the confines of the White town. This could not be otherwise, for most White enterprises have Black employees and most White households Black servants ... [D]omestic servants "live in" in the White areas, being provided with separate rooms in the backyards of their employers' houses ... more Black people are at work in White areas during the day, but housed in the location [township] at night. ... There is also a flow of casual shoppers or sightseers into the White business areas. Still, even for a living-in domestic servant, the location community is and must remain the context of personal ties. ... The purposes for which the Xhosa enters the White town are formal and instrumental. Curfew laws emphasize his exclusion: no location resident is allowed in White parts of town between 11 p.m. and 5 a.m. without a special pass. Moreover, while Black faces are common enough in town, the appearance of a White face in the locations [townships] is so uncommon as to draw stares.

Over time, economic pressures and changing views on the "proper role" for black African people together and differentially led to the legal acceptance of family (in contrast to single-male dormitory) housing (Beavon, 1982; Reintges, 1992; Maylam, 2001). Townships throughout the country were established, with construction beginning in the well-known site of Orlando in Soweto in 1930. They were typically conceptualized as apart rather than of the city despite being economically linked: townships were generally a distance from existing cities, with undeveloped land in between and thus often not spatially abutting the "white" city. Cook (1992) refers to the formation of Khayelitsha, a township established for African residents of Cape Town, as "a new town," with its own mayor and town council (see also Makhulu, 2015). Townships were provided with limited services, but what finances they received came via the national government, not directly from the nearby "white" cities. The structures of governance for townships varied over time and space, although during the apartheid era these legal structures were never granted powers that would enable any real governance by the residents themselves. Such governance structures were, therefore, generally viewed by residents as delegitimate (Mafeje, 1978; Chaskalson, 1989), like the other forms of partial representation offered by the ruling National Party. Important for my argument here, one of these structures was deemed Bantu Urban Councils (1961); this title explicitly identifies areas of black African residence as *urban* spaces.

Generally speaking, much happened in the following decades and political scholarship focuses on these as the decades of interest. Given the focus of this chapter, however, I skip to the establishment of new administrative boundaries, one of many controversial undertakings of the postapartheid government. Although transitional processes occurred before this, the first national democratic election took place in 1994. The African National Congress (generally understood to be the primary national liberation party, although not the only party to contest white rule or win in the democratic elections) largely supported efforts to create "unicities" to transcend the racial splintering of apartheid planning (Turok and Watson, 2001). The slogan "one city, one tax base" emphasizes that the new administrative units would combine formerly "white" cities and townships and redistribute municipal funding. This process certainly was not without controversy and opposition: Turok (1994), for example, details the efforts to escape incorporation in Sandton, one of the wealthiest neighbourhoods on the continent.

The most recent iteration of legal boundaries in South Africa is the creation of wall-to-wall municipalities, eliminating cities as administrative units and obviating the city as a legal entity (Pycroft, 2000; Cameron, 2004; Hart and Sitas, 2004). The significance of this for both academic analysis and everyday language remains unclear, but as difficult as any definition of city might be, there is surely none that would map onto contemporary municipal boundaries.

I conclude this section by drawing attention to definitions of key urban terms from the South African census as it captures some of the ambivalence in contemporary usages. (A more thorough analysis of policy documents would surely usefully contribute more nuances and potentially add regional variation here.) The census notes four broad settlement types: (1) formal urban areas, (2) informal urban areas, (3) commercial farms and (4) tribal areas and rural informal settlements. Interestingly, despite occurring twenty-seven times in the Metadata explanation of terms (Stats SA, 2011), there is no definition of city; town is similarly frequently used and undefined. Definitions provided in Table 4.1 suggest that city, suburbs and townships are parts of an urban area. Townships are both town and historically beyond the town or city. Despite noting historical terminological

Table 4.1 Definitions from the South African census metadata

Urban area	A continuously built-up area with characteristics such as type of economic activity and land use. Cities, towns, townships, suburbs, etc. are typical urban areas. An urban area is one which was proclaimed as such (i.e., in an urban municipality under the old demarcation) or classified as such during census demarcation by the Geography department of Stats SA, based on their observation of the aerial photographs or on other information
Rural area	Any area that is not classified urban. Rural areas may comprise one or more of the following: tribal areas, commercial farms and informal settlements. (See settlement type)
Township	Usually a town or part of a town. Historically, "township" in South Africa referred to an urban residential area created for black migrant labour, usually beyond the town or city limits. Reference is sometimes made to "black township," "coloured township" and "Indian township," meaning that these settlements were created for these population groups. By contrast, the white population resided in suburbs. Informal synonyms for township are "location," "kasie," "ilogishi." Generally, every town/city has one or several townships associated with it.
Suburb	A residential area within the boundaries of a town or city. Historically, suburb referred to a white residential area.
Metropolitan area	A metropolitan area is a large population centre, consisting of a large metropolis and its adjacent zone of influence or of more than one closely adjoining neighbouring central cities and their zones of influence.

changes, however, the contemporary definition still does not include a township as potentially *being a city* or *of the city*.

I now turn to a series of brief interviews to begin articulating what comes into focus when examining vernacular ideas of city.

Notes on safety, fieldwork and being at ease in South Africa

Over the years, a lot of (particularly white South African) people have been surprised that I regularly went into nonmajority-white areas alone on foot and in mass transit. That does not mean it is a crazy thing to do, but it is true that bad things do/have/will continue to happen. Plenty of international researchers say "I did it and all was fine." That does not mean it is actually fine for everyone.

My experience and what I have seen from interactions with dozens of other international visitors is that much of what makes for successful visits is related to but not reducible to identity: it is in how you carry yourself. People can read whether you are at ease. It is not *being* a racial or national outsider, but *looking like an uneasy outsider* that puts you most at risk. Saying so is not to blame those to whom bad things happen. I put this aside here to prompt a deeper engagement with and enable more reasonable navigation (particularly for students or those new to such work) of the competing claims of "it's fine" and "it's not fine."

Awareness of a way of being reduces risk, but of course, this is no guarantee that nothing will happen! If you spend enough time in higher risk areas, things will probably happen. I have had my bag taken more times than I care to admit, usually when I was distracted. I am fortunate and grateful that the financial implications were frustrating but not deeply impactful and that associated violence was minimal. When, for example, my groceries were once taken out of my hands, I was able to turn around and walk back to the store, and walk home a different way. Mostly, I chalk it up to experience, but also share this section hopeful that the experiences might be helpful for more than just me.

There are people (mostly tall, white men) who walk confidently around the world. They tend to be able to go to areas with higher street crime rates without issue (and in my experience, often without self-awareness about why issues did not arise). I am not a person who can confidently perform that disposition: I am small, curly-haired and have been described as maternal and earthy in appearance. I identify as a white woman, but have been described by global south residents as seeming to fit in well enough from Cape Town to Cusco. This appearance and disposition reportedly tends to make it fairly easy for others to feel at ease with me, and I am aware that it has made it easier for me to walk with ease through many unfamiliar places.

When I have been in South Africa for some time, being at ease across different types of urban spaces happens without much conscious effort. When I have been away, I have developed the habit of intentionally performing my "at ease" disposition. Being aware of what is around without darting eyes. Attending to a heart rate, and working to calm it: if the heart rate is up, others can tell. This also includes wardrobe considerations. For me, a long skirt that is clean but not high quality. But I have long worn such things, and feel at ease wearing them. I used to purposefully not wear a watch, because asking for the time is a common strategy for drawing your attention. This matters a lot less now that everyone is assumed to have a phone! Better to look at your watch then be urged to pull out your phone!

I also found that being at ease does not always translate across different kinds of higher risk spaces. I was, for example, quite disoriented in a higher risk area in a large US American city after many years in South Africa. I could not read the social cues and performed the wrong body language. It is worth being thoughtful in the inverse direction too: being at ease across different spaces in northern cities might not necessarily translate into being at ease in southern ones.

If one can be at ease, I find it increases the odds that one can move around higher risk urban areas with reasonably low key interactions with others. Practicing helps: short visits with others in which one attends to disposition, trying to work towards producing ease. Like with public speaking, if one can act calm, others read calm, and eventually one feels more calm too. But. And this is a big but. If this is too much – if it is early in the process, if difference makes one really uneasy – one might take a research assistant, prepare for complicated encounters or find a different data set.

Methods and modes of response

The conventional structure of a scholarly argument is to identify a gap in the literature; we spend much less time explaining the roots of the recognition of a gap. But literature is not a neat set of objects with spaces between: recognition of gaps must come from somewhere. One of the wider purposes of this book is to argue for better explanations of how we come to our research questions as a tactic for pushing beyond the deployment of existing theory. The questions addressed here are rooted in my nonacademic life as well as the classroom, as described in more detail in Chapter 3. My methodology began with a struggle, and the questions developed here seek to provide data to make better sense of this struggle. I do not have an answer for how we properly address this as a scholarly community when there are only so many words per article, and different papers do different things. We might find creative ways of sharing what is behind our work on blogs and websites, although I am mindful this might equally add uncredited burdens. My hope is that the southern urban critique pushes us to shift the balance a little more towards understanding the roots of research projects and the construction of questions as part of our formal research process rather than thinking of such details as "background."

The methods undertaken for this study were developed with my then PhD student Anesu Makina, who also conducted the fieldwork. In what follows in this chapter, "we" refers to Anesu and me. We first pondered whether we could draw on real-world discourse for our analysis rather than conduct interviews. We considered analyzing newspapers, but while we found plenty of examples of relevant terms (urban, rural, village or such), it was less clear how we could understand the authors' meanings. So, we decided we needed to find a way to get different people to talk about the same locations in order to enable our comparisons. We

were concerned that if we simply asked respondents for descriptions or definitions, they might lean on official or academic usages rather than drawing on their
own vernacular (particularly given Ferguson's, 1999 observations in terms of the
adoption of outdated textbook definitions).

We therefore developed a set of photos as the basis for brief photo-elicitation
interviews (see Harper, 2002; Bates *et al.*, 2017). We chose eleven images of
different built environments, covering what we thought of as common structures
across the country. We intentionally selected images that were South African but
neither from the area where we would conduct the research nor of clearly identifiable sites. In the northern summer/southern winter of 2017, Anesu returned
to South Africa where she conducted brief interviews (2–10 minutes). She asked
respondents whether a picture looked like their idea of city. At the end of the series
of images, Anesu asked a single open-ended question: what, then, do you mean
by "city"?

The research was conducted in Tshwane Municipality in Gauteng, an area
where Anesu has lived most of her life and where Anesu and I met while I was
working at the University of Pretoria. Tshwane's boundaries include the pre-1994
administrative capital of South Africa, Pretoria. Respondents were recruited in
and near grocery stores in shopping centres and malls in three suburbs: Hatfield
(110 interviews were conducted), where the main campus of the University of
Pretoria is located; Mamelodi (103 interviews), a township 30 km east of the
central business district (CBD) and Brooklyn (only 15 interviews were conducted
due to recruitment issues), a wealthy suburb that was part of the city of Pretoria,
just 5 km southeast of the CBD. We were less successful in an additional site:
despite similar methodology and multiple attempts, no responses were obtained
in a suburb resided in primarily by white, working class Afrikaans speakers (Pretoria North). The vast majority of these residents also speak fluent English, so
language was unlikely an issue here except in that speaking Afrikaans might have
differently positioned the researcher. The majority of respondents we were able
to access are black persons; analytically, then, we group white, Asian and col-
oured participants together, mindful of the limitations of this. At times, Anesu
approached people who were walking with others; some of these small groups of
two to three persons opted to participate in the research together.

Anesu made notes after each interview to capture both responses to the open-
ended question as well as comments made during the photo-elicitation, including
the conversations between respondents during the small group interviews. Given
the brevity of the interviews, we did not ask about age or race, but Anesu noted
her estimates. We also did not enquire about nationality or place of residents, and
therefore explain our findings as perspectives *in* South Africa and *in* certain locations, rather than *of* South African citizens or *of* residents of particular locations.
We are mindful of the many reasons why race is problematic construct, but as our
data suggest, it remains an important social category that shapes urban imaginaries in South Africa. Our data are of course complicated by the strong correlations
between race and location in our sample, and a likelihood of class difference
as well; more work would surely help us to disambiguate these interrelated

demographic influences. Given what we understand to be the salience of race in our findings, neglecting this in our analysis would limit both our understanding of the data and the ongoing impact of race in postapartheid South Africa.

South Africa is a polylingual country, and while most residents speak English, language did create some methodological challenges. In Mamelodi, Anesu was accompanied by a black South African research assistant who helped facilitate interactions. Translations were sometimes provided to help the respondent understand the research. Translation was, however, limited and avoided where possible to retain emphasis on the English word "city" itself; literature, experience and our interviews suggest few terms are directly interchangeable across different languages. Mayer (1961) for example describes a series of different terms used by isiXhosa speakers in and around East London; different words have different connotations and may reference different parts of a city. We are unaware of any further research examining the use of urban vocabularies in African languages, but note that this would undoubtedly be an interesting and worthwhile project.

Given the novelty of our question and approach, we first reflect on two key patterns observed by Anesu to contextualize how respondents sought to answer our questions. Across all three sites, respondents tended to *state the type of area in the image* before categorizing it. This gave us insight into the typologies being deployed by the respondents, and mid-range categories that were then used by the respondent to classify an image as "city" or "not city." For example, respondents were fairly consistent in naming the image of a house in a formerly white residential area as a suburb; they differed in whether a suburb was city or not. Some noted the absence of a fence (we chose this image to enable the respondent to see the house more clearly without the foresight that this would shape the respondents' interpretation!). The absence of a fence, for some respondents, was taken to indicate that the image was of a safe area and thus could not be city; some speculated it might be an image taken in a town, presumably a unit with a smaller population. And yet, similar to my experiences in Pietermaritzburg described at the start of Chapter 3, "town" was also used as a mid-range category when describing the shopfront in a CBD; this image was noted to be in town, and in this case, town was always identified as being "city." In other words, "town" was both an indicator of "smaller than a city" and as a substitute for CBD. The categories developed by the respondents were often based on associations with familiar locations. Many in Mamelodi pointed at a nearby spaza shop before giving an affirmative response to our image of a spaza shop. Others described structures in the images as being found "everywhere" (this happened in response to the images of the township house, the RDP settlement[1], the shopfront picture and the spaza shop), highlighting the types of images that presumably dominated the respondents' own spatial imaginaries.

The second notable tendency is that, if a response was not immediate or unambivalent, respondents often *looked for specific cues in the image* to determine their answer (such as the fence, as described above). For instance, with the image of the formal township house, many identified the category "township" then examined the image for clues to determine if it was an urban or rural township.

Several participants pointed to the presence of pillars on the pavement (sidewalk) as evidence for a "yes" response. One respondent made note of a building right next to the wall which resembled a backroom or second house built quite close to the house in question; for the respondent, this indicated density and also led to a "yes" response. For the image of the RDP settlement, some respondents reported that they gave a "yes" response because of the presence of high mast street lights, while others argued for a "no" response since the RDPs in the respondents' own (rural) villages reportedly had high mast street lights. The flush toilets were also identified by respondents as a reason to have given a "yes" response; respondents argued that rural RDPs do not have this type of flush toilets. For the aerial view image of a city skyline, respondents who gave a 'no' answer typically identified the presence of greenery and a swimming pool as reasons for exclusion.

We raise these points to indicate, firstly, the interest that many respondents had in our questions and the thoughtfulness with which they engaged with the process. It has not always been my experience that respondents enjoy participating in research; it has, at times, felt like a process of extraction to be endured by the participants for an imagined broader good. I do not mean to read too much into this aspect of the research, but was pleasantly surprised that respondents seem to find our questions – what might be taken as an academic exercise in terminology and classification – to be stimulating, thought-provoking and maybe even worthy of their attention. It helped us to feel like we were onto something interesting. Secondly, while I believe we selected our images thoughtfully, we had not expected the respondents to attend so carefully to the details of the images. This complicated our analysis, for while all respondents answered our questions about the same set of images, these images are composites. Were the work conducted again, one might source simple sketches rather than photographs to develop a more consistent baseline before enquiring into details such as fences, infrastructure, backyard houses, etc. We do not believe this takes away from our wider argument, but instead points us to a much wider range of considerations than we had anticipated and provides grounds for future research.

What is the city?

The research participants provided a diversity of answers both to the binary questions and the open-ended prompt. Mindful that our data are not representative of wider populations, we do not perform detailed statistical analyses but instead focus on general trends. The data collected for this research provide some useful insights, but also provoke many questions. Rather than definitive claims, we instead posit possible interpretations and tentative causality, and suggest areas for further enquiry. From the data, we seek to begin articulating the characteristics of the city as narrated by our respondents. We are particularly attentive to the edges of definitions, for they highlight uncertainty and contestation over what is and is not the city. Insights into these edges are drawn as much from observing as research participants grappled with specific questions as the final answers, for these struggles point us to points of dissent and deliberation.

Table 4.2 Combined rankings (most to least identified as city)

Image	Percentage "yes" responses (n = 228)
Aerial view of high rises	90%
Inside of a shopping mall	86%
Shopfront in a central business district	69%
Block of flats (apartment building)	64%
Formal township house	46%
House in a formerly white suburb	46%
A spaza shop (an informal tuckshop found in townships across South Africa).	39%
Cape Dutch style farmhouse	38%
An RDP settlement	32%
Informal settlement	24%
A rural homestead with round and rectangular houses, cows and a fence	4%

These divergent answers indicate that the meaning of the city is fraught: lines of difference are widely shaped by age, experience and residence, inflected but not determined by the crossings of these with race and class. This diversity of responses is complicated by answers to the open-ended question. While far from identical, these responses tended to reinforce the notions described by Ferguson (1999) as those from outdated textbooks, a point we return to below.

Three photographs provided fairly uniform responses across the three interview locations (see Table 4.2). The image that garnered the most "yes" responses was an aerial view of high-rise buildings (90%). This rank order is fairly unsurprising to us as the image shows a CBD with high-rise buildings; we suspect it is the closest of our set to a global imaginary of an urban built environment. The second highest rank, however, was unexpected: an image of the inside of a shopping mall (86%). The photograph taken in a former Bantustan (a territory legally separated from South Africa during the apartheid era as "homelands" for black African people) received the most no responses (4%). This image includes dozens of buildings (both round and rectangular houses) as well as cows and fences. A "no" response here was also as expected, for it captures what we anticipated would be the most unambiguously rural (and thus not urban) scene of the set. The photographs of informal corrugated iron housing and of an RDP house both garnered mostly "no" responses, but as discussed more below, the percentages varied substantially across the different sites where we conducted our interviews. Strikingly, for the remaining six images, the majority answer (yes or no) differed across the suburbs; there is a notable difference in both the rank order and percentages from different suburbs (see Table 4.3).

In the open-ended question (what, then, do you mean by city?), respondents most commonly identified cities as places with tall buildings, good infrastructure and extensive economic activity. Some specifically used the term "modern," including as justification for including the township in the city. Some repeated

Table 4.3 Rankings by suburb (most to least identified as city)

Hatfield (% yes, n = 110)	Mamelodi (% yes, n = 103)	Brooklyn (% yes, n = 15)
Inside of a shopping mall (89%)	Aerial view of high rises (94%)	Aerial view of high rises (93%)
Aerial view of high rises (85%)	Inside of a shopping mall (84%)	Inside of a shopping mall (80%)
Shopfront in a central business district (76%)	Block of flats (80%)	Cape Dutch style farmhouse (53%)
Block of flats (53%)	House in a formerly white residential area (72%)	Block of flats (33%)
Spaza shop (28%)	Shopfront in a central business district (69%)	Formal township house (33%)
Formal township house (26%)	Formal township house (68%)	RDP settlement (33%)
House in a formerly white residential area (24%)	Cape Dutch style farmhouse (55%)	House in a formerly white residential area (27%)
Cape Dutch style farmhouse (20%)	Spaza shop (55%)	Shopfront in a central business district (20%)
RDP settlement (17%)	RDP settlement (47%)	Spaza shop (13%)
Informal settlement (11%)	Informal settlement (42%)	Informal settlement (0%)
Rural homestead (3%)	Rural homestead (7%)	Rural homestead (0%)

the distinctions found in the literature reviewed above: there are rural areas in the city. While most did not, some respondents explicitly described cities as historically white places. Normative words were also frequent in the open-ended section: the city is the nice area; it is a place one wants to be. Even those who described it as "noisy" or a "concrete jungle" did so with positive undertones, a notable contrast from Mayer's (1961) observation of the connotation of such words.

Characteristics commonly identified in the scholarly literature as what makes a city were, at times, considered by our research participants. Respondents thought about density when evaluating whether backyard shacks were present in the image of a township house. Our interpretation is that such investigations were largely ancillary rather than central to respondents' evaluations, evident in the association of dense homesteads with rurality and well-built farmhouses with the city. Economies were also noted, but the emphasis here was on cities as sites of jobs and consumption rather than as centripetal to production or urban economic flows. Industry and material production were never specifically mentioned, possibly in part due to the specific configuration of South African economic geographies in which these are often in urban peripheries or quasi-rural areas (Rogerson, 1974); we are also mindful that this was a gap in our photo array. Infrastructure, in contrast, was often noted; some respondents pointed to formal or informal infrastructure as indicative of urbanity or its absence. This broader picture suggests provocative comparisons with notions of postindustrial cities that are primarily sites of residence, leisure and consumption; what is different here is that respondents positioned these factors as part of the definition of city, and many positioned themselves as workers contributing to or aspiring to participate in this.

Is your city my city? Race, place and the meaning of the city

Respondents in Mamelodi employed the widest definition of city, evident in comparable or higher rates of "yes" answers for all of the images. For example, even though most respondents in all three sites gave a "no" response to the image of an informal settlement, 42% of respondents in Mamelodi gave "yes" answers (in contrast to 11% in Hatfield and 0% in Brooklyn). One respondent specified, "there are places like that here in Mamelodi and Mamelodi is in Pretoria and it's a city" (although new administrative boundaries complicate the legal existence of Pretoria, it unsurprisingly still exists as a city in the local imagination!). We also find this pattern of more "yes" answers from Mamelodi in the image of a formal township house, as well as a house in a formerly white suburban (see Table 4.2).

One additional point of interest in the data from Mamelodi is that age appears to influence the interpretation of whether a township is part of a city. Of those who responded "yes" for the formal township house, 19% were older, 35% of people were middle aged and 45% were younger. While the sample is small and ages visually estimated, the data point us towards the importance of history and changing usages of urban vocabulary: younger people in the township are more likely to claim that the township is (of) the city. This pattern, however, is not evident across our research sites, as the majority of respondents in Hatfield were young and provided a lower percentage of "yes" responses. Further, respondents in Hatfield who volunteered that they lived (presently or in the past) in townships did not necessarily give "yes" answers for the township house or informal settlement.

Given our difficulties in garnering participation from white potential respondents, we are cautious in our reporting around race. This difficulty is in itself a useful reminder of the salience of race in and beyond research. Because racial differences were central to our understandings of the variances in urban vocabulary prior to conducting this work, and the limited results do indicate strong trends, we include some indications from our data that point at ways in which notions of the city continue to be racialized.

Two findings stand out in our analysis: while the results vary across suburbs, there is a notable difference in the results by race to the image of a house in a formerly white suburb. Fifty-one percent of black African research participants responded "yes," compared to only 15% of the other respondents. Further, some respondents, in naming the area, explicitly categorized the image as a "white people area," "white people suburbs" or "esilungu" (place of white people). This reinforces the notion that, for many black people in South Africa, "city" is a term used for areas that were previously proclaimed to be white areas, including but not limited to white residential suburbs. The second image that received a particularly racialized response was the image of a Cape Dutch style farmhouse. A total of two non-black African respondents across all the sites gave a "yes" response, in contrast with 67% in Brooklyn and 42% of the total number of black African respondents. Some black African respondents explicitly identified the house as "a white people house," "Afrikaner farm," or "white people's rural areas."

By and large, respondents asserted that areas that were recognizably formerly white areas are part of the city. This association was most explicit for older residents who likely had first-hand experiences of the exclusive regulations at the height of the apartheid state. For urban researchers familiar with South Africa, it is surely no surprise that such a significant part of apartheid policy continues to have impacts. And yet, as scholars increasingly draw on international theoretical apparatus and participate in international discussions (and the apartheid era is increasingly seen as an historical phenomenon), the impact of the association between race and the city may well be increasingly obscured.

Is the city for me? The aspiration to belong to the thing-called-city

A desire to belong, to identify where one is as part of "city," is evident in our data in a way that we had not anticipated. Specifically, respondents appeared to, at times consciously, extend their definition of city to include themselves. Respondents in Mamelodi in particular often compared images with their immediate surroundings, gesturing that this place too is "city." If city is modern, nice, good and desirable, then this is a vision to which respondents not only wanted to belong, they asserted their belonging. Provocatively, not all respondents wanted such an inclusive definition. Those closer to the urban core, the historically white city, regularly asserted narrower interpretations that excluded others (poorer, more peripheral) in their narration of "city."

Our data provide but a limited entry point into understanding vocabulary and imaginaries of the city in South Africa. A single study surely cannot begin to address all aspects of this question, let alone its impact for urban studies that take the southern urban critique seriously. Our approach emphasized the built environment rather than, for example, economic or social relationships, and this surely obscures some of the key associations for ordinary residents. And yet, even in this emphasis, normative judgments about the city have come through: for many research participants, the city is a place you want to be, be part of and to which you want to belong.

Implications and future directions

The last two chapters have engaged with historical and contemporary literature as well as contemporary everyday language as tactics for analyzing the entanglement of European, colonial and apartheid ideas with race, modernity and the city. The words used by residents to narrate what is (not) city demonstrate fraught associations, ambiguous polylingual translations, racialized values and troublesome political and economic tensions. Our interviews raise questions about the impact of positionality and experience on ideas of city. They also provide some insights into specific problematics people think through when seeking to determine whether a space is (not) city.

There are both practical and political implications of these findings for urban research. Again, what one predetermines as "city" or "not city" will surely shape urban analysis as it shapes what empirical phenomena are included in a study. The elimination of "cities" as legal units in South Africa removes any easy-to-adopt common boundary, and means that urban researchers will need to develop new units for analysis. In my own first experience working on an international *urban* research project, we simply adopted the new municipal boundaries: as a result, our research sites included "periurban" places that looked very much like the image that respondents clearly identified as "not city" (see Goebel, 2015). Different meanings will also impact any work that includes conversations with research participants: if the researcher and the research participants are actually using the terms differently, without realizing it, this will surely negatively impact the veracity of the work (see Pierce and Lawhon, 2016).

The associations of city with positionality, experience, and racialized modern spaces, as well as being a site of (aspirational) belonging, give weight to wider concerns about not only how we operationalize the urban in research, but also how ideas of the city are incorporated in allegedly progressive efforts. If a city remains a (colonial/apartheid) white city in people's imaginations, then what does that mean when residents are asked to participate in *city* visioning exercises (see Pieterse, 2002; Parnell and Robinson, 2006; Marx, 2011; see also Myers, 2015; Battersby, 2017)? While researchers have examined participation in many urban planning processes, they have largely done so from an assumption that the city is a material, spatial phenomenon rather than as a historical, power-laden idea (again, see Sayer, 1984). While the purpose of such exercises is often meant to open planner's and residents' imaginations, might the use of the word city itself unconsciously shape and constrain imaginations? And while I have no evidence of this being a conscious or intentional motive, nor have I seen this observation in the scholarly or grey literature, it is quite possible that the obviation of the city as an administrative unit in South Africa is informed by this fraught association: one way to overcome the whiteness of "city" is to simply erase it as a political unit. Reading this action through an economic lens as a strategy through which to redistribute resources, as is the common explanation, may be both accurate and overly reductive when understood in this wider context.

The findings analyzed in this paper also raise deep questions about the relevance of political claims regarding ongoing discussions over the "right to the city" (e.g., Parnell and Pieterse, 2010; Huchzermeyer, 2014; Goebel, 2015; Makhulu, 2015). Without discounting the fact that the term has been deployed by South African social movements (and see Chapter 8 on environmental justice for some concerns with outward-facing social movement vocabulary), what might such a claim mean for those who see themselves as excluded from its definition? Does the liberatory potential of the concept wane if "city" is a claim to formerly-white-space, to formal modern infrastructure? Or, we might look to the youth of Mamelodi from our interviews to provoke an alternative frame: a right to be *called* city. Such a frame is more than a semantic reworking, but includes a reevaluation of the normative judgments inherent in the centuries of urban exclusion and exclusion from being called urban, a point I return to in Chapter 9. One way through which to take forward research might be to interrogate whether the assumed (modern, formal) city really is the ideal to which residents aspire. Here my concerns dovetail neatly with Robinson's (2006) urgings to open urban theory to more cosmopolitan understandings of what it means to be urban. I wonder, however, whether such exercises might either directly seek to unpack urban vocabulary, or be conceptualized through different vocabularies.

In the epigraph for this chapter, drawn from one of the interviews on which our research is based, a young man near the University of Pretoria succinctly asked a question that has for too long been ignored in urban studies: who is to say what the city is? I have intentionally paired his question with an off-the-cuff observation from a resident of Soweto, several years older, made several years beforehand, that being called a city is in itself a compliment. This chapter seeks to make deeper sense of these two observations in relation to each other, for limited attention has been given to the politicized association between what people understand city to be and its impact on how people live in, understand, govern, re-create and aspire to be (in/of) the city. My point here is not that South Africans are somehow "wrong" for making this conflation, and merely need to adopt the scholarly language that no longer explicitly retains these associations. I am urging a deeper recognition that South African cities – the bricks and mortar, the plans, the rules, the social processes that happen within them – are and continue to be underpinned by *ideas of what a city is*. These ideas are connected to global ideas, but also have their own unique and situated history.

More broadly, then, my argument is that the answer to the urban question – the generally understood boundings of the city in urban studies – was developed from ideas of and about sociomaterial cities of the global north. It is not that this consequent scholarly idea of the city cannot be deployed elsewhere: indeed, as shown in Chapter 3, it often has been, and this has yielded useful insights about cities globally (see Hannerz, 1980). Here, my concern is with the entanglement of academic theory with its geographical roots, including the assumption that approaches derived from ideas about northern cities are scholarly, while those derived from ideas of cities elsewhere are not. I suggest the need for answers to the urban question to emerge from a more pluralist range of sites and sources, and that these can all usefully contribute to developing a more spatially plural, postcolonial understanding of the urban question.

Note

1 RDP stands for Reconstruction and Development Program, a national policy framework adopted in 1994. While the program no longer exists, state housing developments of small private homes are typically still called "RDP houses" in the vernacular.

5   Teaching urban geography after the southern urban critique

With Lené Le Roux

I start this chapter with a pause to take stock of where we are in the wider arc and argument of the book. I began this book by constructing an argument for postcolonial unlearning and learning anew, mindful that this is not the only grounding upon which scholars have asserted a need to provincialize theory and include southern cities in urban studies (see Chapter 2). In the last two chapters, I argued that even the most foundational and fundamental question of urban studies – what is a city? – requires postcolonial attention. In this chapter, I work to move the conversation forward by asking: what are those convinced of the southern urban critique to do with these insights? What do these scholars want urban studies to look like? And how might such scholars work towards enacting it?

In the next two chapters, I articulate possible pathways after the southern urban critique. Unlike the various propositions discussed in Chapter 2, the pathways started to be imagined here, and fleshed out in more detail in Chapter 6, do not hold the potential to be additive (although it is plausible that an eclectic urban studies might well contain advocates for and followers of different pathways). These different pathways call on us to make difficult choices on two particular fronts that are interwoven, but useful to articulate separately. The first is strategic, and is about convincing the wider urban studies community to attend to the southern urban critique. It is a question of what scholars ought to do now in response to our historical moment. The second is about how scholars conceive the ability to theorize, including questions about the geographical travellings of theory. In this chapter, I use the example of teaching and textbooks as a lens through which to begin teasing out what divergent approaches to urban geography might entail. This includes both epistemological questions as well as practical ones, for while we might hypothetically ask researchers to read everything and speak to everyone (see Sheppard *et al.*, 2013), it seems easier to prudently imagine constraints when naming a course and developing its syllabus and reading list.

Of course, the challenges for teaching are not the same as those for an intellectual community. I therefore use coursework as a starting point in this chapter, and follow it in Chapter 6 with an investigation and articulation of pathways for the structure of urban studies as a field of academic enquiry and as an intellectual community. Throughout both chapters, I seek to demonstrate the strengths and weaknesses of different approaches. The difference between the pathways may well seem marginal and have minimal impact on everyday academic practice for many scholars. Yet these pathways go to the heart of the implications and endpoints of the southern urban critique as an academic-political project. These pathways offer fundamentally different outcomes of research across cities in terms of what and how knowledge travels, what and how we generalize and what and how we ultimately can know.

I start here with a brief examination of how southern cities are typically represented in urban geography textbooks, mindful that this is a proxy and surely neither a perfect measure of how urban geography is taught nor of how the sub-discipline is conceptualized (see Myers, 2001 or Sidaway and Hall's, 2018 special issue for more on global concerns with textbooks). With help from then PhD student and research assistant Lené Le Roux (and in this chapter, "we" refers to Lené and I), we analyzed the structure, content and representation of cities in five urban geography textbooks (urban geography provided a more focused set than urban studies more widely; the textbooks analyzed were Hall, 2006; Pacione, 2009; Jonas *et al.*, 2015; Kaplan *et al.*, 2014 and Knox and McCarthy, 2012 which we believe are the most commonly used books on the market[1]). More detail on the methodology used for the textbook analysis, as well as more details about its findings, can be found in Lawhon and Le Roux (2019). This chapter draws from the more detailed study with a focus on the implications of the different structures, content and representations.

This work is motivated and informed by my experiences teaching at several South African universities (the Universities of KwaZulu-Natal, Cape Town and Pretoria), and in particular teaching urban geography at the University of Pretoria. For the idea of studying textbooks themselves I am (yet again) indebted to Garth Myers, whose 2001 piece on representations of Africa in human geography textbooks provided a tactic through which to investigate wider questions. The chapter is also informed by teaching development geography in the United States and discussions with Joe Pierce about teaching urban geography in the United States. When teaching urban geography at University of Pretoria, my students and I struggled to develop common vocabulary through which to talk about "the urban" (see Chapter 4). As is often the case when starting a new job, I inherited the course and a textbook, but somewhat atypically, I literally took over the course midway through. It is worth saying strongly that my intent here is not to criticize my colleagues, who I believe were well-intentioned and performing a task similarly to many others globally. I taught from Pacione (2009), (which has a separate section on "Third World Cities" at the back of the textbook) and supplemented the chapters with South African newspaper articles that sought to provide local examples of the theme. My struggles with the wider framing of the book as well as to make the textbook relevant provided the motivation for this investigation; this is but one small step towards a long-term ambition of making it easier to teach differently.

Representation of northern and southern cities in urban geography textbooks

Unsurprisingly, the urban geography textbooks we analyzed focus primarily on cities in the global north. Hall (2006) stands out as having minimal coverage of southern cities (an historical analysis that looks at previous editions might be of interest here; for now we flag that urban geography is changing!). Kaplan *et al.* (2014), Pacione (2009) and Knox and McCarthy (2012) have separate sections dedicated to the global south, although southern examples were occasionally found in other sections of these three texts. Roughly one third of the content of these texts addresses southern cities (a notable portion of this is historical information, as discussed more below). We call this predominant structure "southern urbanism" (although the textbooks themselves do not explain a particular reason for their practices, nor use this term): southern cities are treated as distinct from northern ones, empirically and theoretically. The inclusion of southern cities in most textbooks makes clear that these cities are part of urban geography, although they represent a much smaller part.

Jonas *et al.* (2015) is the most recent text we analyzed, and the only one that is explicitly critical; it has international authorship, and these factors likely contribute to the fact that it stands apart from the other texts. In contrast with the other texts, this textbook integrated material about the global south into chapters with wider themes. We broadly categorize this approach with the dual label "global urban studies/world of cities" (which I work through in more detail in Chapter 6), drawing here on two terms both used by Robinson (2005, 2016). We interrogate the extent to which this text adequately represents the southern urban critique below.

The kind of information provided about northern and southern cities follows a distinctive pattern. The core themes presented in sections on the global north cover a greater range issues and are more analytical (e.g., city formation, economic, infrastructure, housing, social and cultural and political). Images of cities in the global north often include large infrastructure projects such as bridges and ports, residential typologies and panoramic high-rise cityscapes. Architecture and urban design techniques, extravagant public squares, malls, street art and graffiti, as well as gentrification were also found. In contrast, individual stories of ordinary lives are more prominent in sections on the global south. This point is well-captured in our analysis of the images in the textbooks: photographs in northern cities are typically of groups or taken at a distance, particularly for images of homeless or destitute urban dwellers. In southern cities, front view, close-up shots of individuals are often present, particularly for homeless or destitute urban dwellers. This parallels wider moves in development studies to focus on everyday processes and practices (see Rigg, 2004; Williams *et al.*, 2009).

Discussions of southern cities focus primarily on the confluence of poverty and informality, and the nexus of human and environmental health. As with our observation of a focus on individuals and the everyday, this is largely in accord with Robinson's (2006) characterization that southern cities are approached through the lens of development. Typical imagery of the global south includes ancient city ruins, contrasts of colonial built environments with indigenous and informal built environments, busy streets and dense public space, street markets and adult traders, traffic congestion, slums and high-rise apartment blocks, refugee camps, homelessness, infrastructure problems in housing, waste and sewerage systems and environmental problems such as smog and flooding. Discussions of southern urban slums bring together these foci in discussions of alternative household utility connections, environmental and health hazards, "extra-legal" housing, tenantship and illegal subdivisions. Rural-urban migration is often linked to slums, framed as problematically resulting in so-called "mega urban regions" and "mushrooming," especially in Asia.

The global north is often used as a reference point to which the global south is compared; the inverse is notably absent from all the texts. This point, again, is in keeping with concerns raised through the southern urban critique. In such comparisons, the global south is framed as being behind or catching up to the north, as well as lacking or as a comparative failure. While surely there are similarities between what has happened in the north historically and the south at present, such a representation places the global south as out of sync (see Chakrabarty, 2000).

Several textbooks have chapters on the history of urbanization and a global map of urban origins, all in the *global south*, although this spatial observation is left unnoted. Equally without comment, the texts then give a historical commentary on *European* cities, providing no justification for this spatial shift (although two of the textbooks have, in their introduction, noted a focus on the United States; Pacione is explicitly global in its title and introduction). While four of the five textbooks do include information about colonialism, they engage the process rather differently. Considerations of colonialism in most texts are largely confined to their negative impacts on the global south (rather than the impacts of these relations in the north); again Jonas *et al.* stands out as the one textbook that does relate contemporary geopolitical and economic relationships to historical colonial relationships, and also considers the impact on indigenous ways of knowing.

Explicit attention to calls for greater spatial plurality in urban geography is limited. Only Jonas *et al.* (2015) directly engage the work of Robinson, Roy, Pieterse and Simone (p. 44–51) in any detail, and do so early in the text rather than seemingly added on as an afterthought. But their presentation makes it easy to misread the southern urban critique as a singular perspective arguing for what we have called southern urbanism, an observation that southern cities are different. This is particularly of interest as the textbook itself is the only one to not separate southern cities into distinct chapters. The argument for global urban studies and the call developed through the southern urban critique to rethink the ways in which we create urban theory are both absent. For example, the challenges to urban theory raised by southern scholars are framed as being a result of increased urbanization and connection rather than a problematic tendency in urban geography at large. Again, see Lawhon and Le Roux (2019) for a more detailed analysis. More generally, and rather understandably, there is no attention given to the problem of how best to include consideration of southern cities. Although both the southern urbanist and global urban/world of cities approaches can be read into various texts, their distinctions are not explicitly identified; again, this feels fair given that the wider literature has given little attention to this distinction.

Implications for teaching urban geography

There is no unanimous or easy answer to the question of how to structure and represent southern cities in urban geography textbooks (and this point is true for the field more widely as well, as discussed further in Chapter 6). There are, no doubt, empirical differences in the prevalence of certain phenomena across many northern and southern cities, and, indeed, part of the critique from southern scholars is the uncareful application of theories developed in the north that inadequately account for these differences. But there is equally no doubt that a north-south binary is an imperfect articulation, and southern urban scholars have also noted concerns with the reification of difference, a kind of difference embodied in the separation of southern cities into their own sections generally at the end of the textbook (see Chapter 1).

Therefore, in this section, we focus not on reconciling whether southern cities should be included with northern ones or treated separately, or on whether cities ought to all be presented in the same narrative and stylistic manner. Our focus here is instead on exploring the implications of different strategies with the hope of prompting further discussion on the relative merits of these strategies. Finally, we discuss our strongest recommendation: that textbooks be more explicit about both the importance and difficulty of studying the plurality of cities.

There are very pragmatic concerns with increasing the scope of urban geography by simply adding more text on southern cities. While textbooks can be longer (e.g., Pacione, 2009 is 736 pages!), this then defers decision-making and puts the impetus on individual instructors to determine what is and is not taught. Realistically, including more about southern cities into the urban geography curriculum requires either *teaching less of the old* to make room for the new or *separating the south* into a course which might parallel, complement or replace urban geography (see Bunnell and Maringanti, 2010; Kenna, 2017).

There is no doubt merit to the teaching of urban geography differently in different places, and surely this is already ongoing. Examples stick best when students can connect, and therefore it is generally useful to include local examples. We appreciated the explicit statements about the locational focus in Knox and McCarthy (2012) and Hall (2006). And yet, when the global production of course materials is so centralized, it may well be difficult for instructors at southern universities not to adopt the readily available texts that are presented as universal and authoritative (particularly if they want to enable students to pursue graduate study elsewhere).

If we work from a foundation that teaching ought to draw on and develop understandings of local contexts (a practice that likely already exists), we must recognize that, at present, there are fewer publicly available materials for creation of an urban geography course in the global south. The Parnell and Oldfield (2014) handbook and Miraftab and Kudva (2014) reader on cities of the global south provide useful starting points for explicitly urban studies courses on the global south (i.e., southern urbanism), but do not substitute as holistic introductions to urban geography. This prompts us to consider whether there are ways to facilitate learning across contexts outside of the production textbooks through international publishers. Examples such as the Association of African Planning Schools (AAPS) (https://africanplanningschools.org.za/) for urban planning might well provide a useful illustration of the kinds of materials that can be locally based but widely shared. The AAPS has, amongst other things, developed Curriculum Development Toolkits focusing on thematic areas with African-country-specific case studies. Urban geographers teaching in the global south might well usefully develop similar strategies for sharing materials, improving the quality of materials and reducing individual effort.

How ought we teach about elsewhere?

There is no easy way to represent and teach about difference; teaching about difference surely makes us and our students uneasy. Scholars have argued that increasing attention to the everyday in development geography can help to counter the "othering" so prevalent in the media as well as the very real physical distances, and this approach can be found in two recent development geography textbooks (Rigg, 2004 and Williams *et al.*, 2009; see also Sparke, 2018). And yet, when the everyday is focal in southern cases but not paralleled with similar enquiry into northern cases, this can ironically be estranging because of its contrast with the presentation of northern cities. The close-up images which we found to be more common in photos of southern cities (see Lawhon and Le Roux, 2019), for example, can be read as usefully closing the distance between the reader and the other, but equally so as an invasion or taking that would not be tolerated of and by northern urban residents.

We suggest that one strategy for thinking through this problematic is to begin with a more careful interrogation of assumptions about how well undergraduate students understand their own hemispheric context, let alone the lives of their immediate others. Most of the students I have encountered in my teaching in five universities in South Africa and the United States have not travelled outside their home region except to visit tourist destinations, and most have not left their home country. Further, their own everyday rhythms provide them with a limited scope of the city in which they live; typically they have moved from being a student in a family home to being a university student. This trend is less true for graduate students, but the wider concerns here are largely applicable for graduate education as well. The arguments raised by development geographers about the utility of everyday narratives for enlivening our understanding of others might well provide a productive lens through which to engage with cities for any audience with limited personal experience. In such cases, in the spirit of comparative urbanism, similar methodologies and presentation styles might usefully draw attention to the diversity of global urban life.

And yet, while this proposal for a mode of enacting a global urban studies/world of cities approach provides a fairly easy antidote, we are cautious about what its impacts might be and hesitant to recommend it in our contemporary historical moment. A factual presentation of empirical differences in a similar style cannot obviate the lens through which students of urban geography read narratives. Were students encountering texts as "blank slates," without preconceived ideas, then this approach might well be sufficient to introduce them to a global urban diversity of, for example, urban diets, pastimes or housing. But of course, students do not enter the classroom without notions of the world. Presenting information without contextualization about how we see "the other" might well mean that urban geography textbooks reinforce problematic notions of difference.

It might feel like too much to ask of urban geographers who write textbooks and teach urban geography to be familiar with postcolonial and development theory. And yet, we suggest that greater attention to this is necessary if urban geography courses are to adequately teach about more than half of the world's cities and urban residents. If we are not willing to require this of the authors of textbooks, then is it better to simply write the south out of urban geography, to separate it off as its own course?

In a context in which there are no easy answers, we propose here a tactic that responds to existing limitations rather than a long-term resolution (more on this in Chapter 6). A productive starting point is to urge textbooks and instructors to address issues of representation and incorporation in the teaching of urban geography. We realize that, from our experience, the majority of students would prefer straightforward claims, definitions, and answers. We also often wish the world was better accounted for in this way! But the poststructural influence of the mid-1990s has helped to lay the groundwork for textbooks not as authoritative truths but as mediated social constructs, and representation as always imperfect. Textbooks often have introductions that explain their framework, audience, scope and limitations, and these should better address the concerns raised above. Additionally, in the main text itself, we argue for presenting uncertainty, naming the difficulties associated with representation and explicitly attending to the problem of how to incorporate southern cities into urban geography. We believe this is useful both for students' engagement with the global south as well as for their understanding textbooks not as authoritative texts but as difficult, temporary representations that tell important, thoughtful, but always imperfect stories.

Note

1 We subsequently discovered an oversight here: we did not use the more recent version of Hall and Barett (2012), which turns out to have also had a new edition (2018) since we conducted the empirical work. We are delighted to see this new version explicitly addresses the southern urban critique, although do not include analysis of it here! We apologize to the authors for our oversight and believe that the analysis that follows continues to usefully demonstrate patterns and concerns.

6 Pathways forward

Urban studies after the southern urban critique

With Yaffa Truelove

Here I build on the analysis particularly in Chapters 2 and 5 to work through more specific pathways for urban studies as a field of scholarly enquiry. This chapter draws on the second half of Lawhon and Truelove (the first half is the foundation of Chapter 2). As before, in this chapter "we" refers to Yaffa and I. In Chapter 2, we demonstrated that scholars making the southern urban critique base their arguments on three primary claims: (1) the south is empirically different, (2) the south has different intellectual and vernacular traditions and (3) postcolonial relations require us to examine the production of all knowledge. Articulating the foundations of the southern urban critique is useful for deepening our understanding of the urban studies we seek, but here we work to more specifically address the range of possible outcomes.

In Chapter 5, through an examination of five contemporary textbooks, Lené and I suggested that southern cities are incorporated into the teaching of urban geography in two key ways: as separate chapters and integrated into thematic chapters. Drawing on the wider literature, I call these approaches, respectively, "southern urbanism" and "a global urban studies/a world of cities." In the analysis following my descriptions of the textbooks, Lené and I wrestled with the difficulties of including postcolonial concerns with representation of southern cities, and the difficulties of asking undergraduate and graduate students to think about unfamiliar places. Mindful that urban geography is but one approach to urban studies, here I work through how the insights drawn from Chapters 2 and 5 can be used to better understand possible pathways forward for urban studies after the southern urban critique, including a more explicit articulation of the conceptual underpinnings of different modes of incorporation.

Pathway 1: southern urbanism is and ought to be studied as a distinct phenomenon

In this mode, the study of southern cities would become (or, to some extent, remain as) a distinctive field. This mode is intellectually compatible with, although does not require, further partitioning into regional studies (African cities, Asian cities, North American cities and so on). It does not preclude analysis of the relationality between the south and the north (e.g., examination of colonialism and/or dependency). The foundation for this call is usually empirical difference, but it is not incompatible with arguments for diverse traditions (e.g., one might argue that knowledge of African cities applies only to African cities) and attending to underlying assumptions (e.g., one might argue that only those in and of the south can truly understand it).

One can undoubtedly find examples of the study of southern cities as distinct, although reading off an ideal mode of urban studies from any specific work is surely imprudent (see Chapter 2). Schindler (2017), for example, is the strongest we have seen articulating a separate "paradigm" for southern urbanism, while Watson (2009) makes a strong case for urban planning through a "southern" perspective. There is another reading, however, and it is unclear to us the extent to which authors writing in this vein believe such categories are "real" or "temporary and to be transcended." Myers' (2011) monograph *African Cities* follows a long-standing effort to define and describe this category as distinct (see also O'Connor, 1983). And yet, Myers queries what urban studies would look like if Lusaka were the postmetropolis, methodologically paralleling Soja's classic urban work. His 1994 work detailed in Chapter 1 as well as more recent work on north-south comparison (Myers, 2014, In press) make clear that Myers is interested in using concepts developed from African cities to inform a wider field of enquiry beyond African cities. Doing so challenges any simplistic notions of African studies being relevant only to or for African cities and calls into question, for our purposes here, what the ideal mode of urban studies of the future might be. Are such articulations actually in support of southern or regional urbanism as a distinct field?

If not intended to support long-term regionalist urban studies, of what purpose are regionally focused articulations, then? We might see them, like the term "global south" itself, as part of a "strategic essentialism," a way through which to garner scholarly attention through emphasizing similarities (see Roy, 2014). Here, the long-term aim is to reduce rather than reify the significance of such categories. Ramogale (2019) makes this point more broadly, in contrast with the Afrocentric frame asserted in Sanders (1992) reviewed in Chapter 1. Ramogale argues, "in progressive thought Afrocentrism is never the end; it is rather a means, a stepping stone to a global platform where scholars engage in intellectual and cultural exchange as equals." As this global platform has yet to be achieved in urban studies, regionalist work at many scales may well be understood as a means towards a different, distant and elusive end. Here, however, we note the importance of not conflating the means and ends and urge scholars writing in this vein towards mindfulness as well as a more specific articulation of the long-term ambitions of the southern urban critique.

Pathway 2: the southern urban critique is an ontological position against universality and asserting the subjectivity and locatedness of all theory

In this articulation, the south is considered to be a source of a wider critique of the production of knowledge. We might refer to this as a postcolonial world-of-cities approach. For example, Radhakrishnan (2008) uses the idea of south as a "concept-metaphor" by which new urban theory can emerge that reveals the "provincialism of dominance," working against theoretical orientations that view the north as universal and south as particular/exceptional. In a similar vein, Roy and Ong (2011) propose that the south serve as a signifier of the locatedness of authoritative knowledge (Theory with a capital T), provincializing knowledge domains normatively perceived as universal. The essence of this concern is that all theory is located, and that to undertake research across contexts, we must attend to the locatedness of the researcher and the theory (including concepts, causality and norms).

This proposition is importantly distinct from the null claim addressed in Chapter 2. It is *not* a claim for particularism and denies a binary between the universal and particular; it is an assertion that there is a middle space between these. One might draw, here, on Chakrabarty's urging for investigations between History I and History II. Here, History I is the more well-known story of capitalism: a story abstracted from Marx's analysis of a particular location. History II is the real world empirical stories, neither Other nor subsumed by History I. History II has elements that do not, at present, mirror the conditions described as conducive to capitalism. The interaction between the allegedly analytical/universal (History I) and the located/lived (History II) produces a middle ground for exploration between particularism and universalism. Roy and Ong (2011) usefully propose a methodological and conceptual approach of "worlding" as a lens through which to apply such locatedness in examining cities globally. For example, McFarlane *et al.* (2017) propose theorizing diverse urbanisms within cities (*intra* urban comparison) in conjunction with thinking comparatively and relationally across cities (*inter* urban comparison) in order to reconceptualize urban politics more broadly. Urban scholars are thus actively seeking to more rigorously understand what it might mean to, for example, generalize without universalizing, develop "mid-level" theory (Roy, 2005), or study multiplicities across differing scales (Robinson, 2011; Schindler, 2014; McFarlane *et al.*, 2017).

While this position can be found in urban studies, our reading is that its precise implications for an appropriate mode of urban studies are still being worked out. Scholars articulating this position simultaneously seek to displace the universal/exception dichotomy that has plagued urban studies while clearly arguing that theories developed in places have relevance elsewhere (Robinson, 2006, 2016; Roy, 2009). We agree with many working in this vein that how exactly we acknowledge and respect the conceptual and theoretical offerings and limitations of this pathway (Robinson, 2016), as well as precisely what aspects of theorization can be made to travel (and to what extent) (see Lawhon *et al.*, 2016), is a key subject for scholarly enquiry.

Pathway 3: the southern urban critique is a tactical argument to decentre urban theory. Having done so, now we ought to work towards general, global urban theory

In this mode of urban studies, theorization that generalizes and explains widely across empirical differences is both possible and useful; even when developed from place, theory can be unlocated. We might refer to this as global urban studies. Here, the point of theorization is to explain, and relate, across difference. The southern urban critique is necessary because southern cities are understudied, and northern theory needs to be provincialized (rather than assumed universally relevant). However, while empirical difference is fundamentally significant, it does not require ontological recalibration. While we do not find explicit articulations of such claims, there is evidence of attempts to navigate this murky water. Specifically, this involves questions as to whether we can now integrate southern cities into more unified modes (and methods) of urban studies, generalize relationally across sites and space, and what that means for the possibilities of urban theory.

Moves to develop a revitalized comparative urbanism offer a productive lens through which to examine ideas about future modes of urban studies. In these, we find authors grappling with what exactly is possible from such comparisons, including a thoughtful reflection on the possibilities of differing forms of universalization (as modes of enquiry or conceptualizations, rather than universal causal explanations; see Hart (2018) for a deeper engagement with different modes of postcolonial comparison). For example, Leitner and Sheppard (2016: 231) note, in agreement with Robinson (2014), "comparative urban research must be undertaken in ways that avoid reaffirming this universalism of the dominant as the implied standard of measure." For us, the utility of this quotation is that it leaves open a question of the possibility of certain types of universalization: surely, there is consensus amongst scholars of the southern urban critique that there are significant limitations to the "universalism of the dominant." But it is not entirely clear as to whether this is a rejection of universalisms of all types (including methodological and theoretical modes of enquiry), let alone generalization and explanation across a wide range of cities. One might here grapple with the implications of Chakrabarty's History I and its utility: for while Chakrabarty is clear that History I is not directly universally applicable or explanatory, and is transformed through interactions with History II, we may read his argument as not opposed to the articulation of History I (e.g., Marx does have utility for understanding India, we just must examine how our understanding of capitalism changes through this case). Instead, any universal articulation of History I ought to be understood as a "place-holder" (Chakrabarty, 2000: 70); History I is dynamic and changing as it is modified through empirical studies. For our analysis here, this means that universals are allowed to be modified through interactions with global south urban studies.

Significant for our analysis here, explicit reference to "deep postcolonial theory" has long been widely absent from methodological reflections on urban comparison (e.g., Abu-Lughod, 1975; Robinson, 2016). Strands of deep postcolonial theory are not necessarily incompatible with generalization, and specifically the mode of comparison, but does trouble the ease with which we can undertake this. Problematic comparisons have been widely accepted as foundational to the modern colonial rationality and are the source of narratives of southern exceptionalism or failure (Hart, 2018). More recently, we can see that urban comparison in practice still requires us to unpack our universalizing assumptions. This concern is exemplified by recent debates on northern-derived conceptualizations of gentrification and its applicability and comparative purchase elsewhere (for example, see Lees (2012) articulation based on urban comparison). Ghertner (2015), for instance, argues that the gentrification concept derives from a set of assumptions and northern contexts that do not hold true globally. This debate is illustrative of a wider concern that comparison can, but does not require, be a careful unpacking of the assumptions that underpin theorization. We argue that before we can understand the wider potential for generalization, more emphasis is needed on the process of unlearning (e.g., identifying core assumptions and evaluating their relevance before generalizing).

This chapter has worked to articulate possible pathways for creating a more spatially diverse and theoretically robust urban studies: (1) southern (or regional) urbanism as a distinct field, (2) urban studies that recognize all theory as partial and located, or what we might call a postcolonial world-of-cities approach and (3) a global urban studies in which all cities equally contribute to knowledge production. While we find much use in the articulation of these three positions, we are concerned that these are largely conceived of as discrete positions (even if rarely articulated as such) with limited possibilities for reconciliation. Yet in working out this argument drawing from the burgeoning literature, we became less clear that answers can be derived through the existing frameworks through which this conversation is occurring.

In the next chapter, therefore, I work through a more detailed encounter with theory itself. Before answering whether a particular "theory" may be, for example, southern, located, general or universal, I suggest we might benefit from first "unbundling" the notion of "theory." Drawing on the wider southern urban critique as well as my own scholarship, in the remaining chapters of this book I seek to identify the particular components being investigated, and from here more carefully specify different components being critiqued. For, as exemplified in the next chapter through the case of gentrification, whether one agrees that there is a universal empirical *phenomenon* of displacement is not the same as agreeing to the universality of causal *explanations*. Nor is it to agree about the *normative implications* and *tactics for addressing* gentrification, including what the goals of action ought to be.

7 Unbundling urban theory

Of what is theory made?

With Lené Le Roux

The notion that theory from elsewhere might retain utility in our effort to understand southern cities is broadly agreed upon by the scholars whose work has been reviewed (and contributed to the argument developed) in the previous chapters (see also Said, 1983; Clifford, 1989 explicitly on travelling theory). Throughout my own work in the previous chapters and elsewhere, I have grappled with what precisely this means in practice in urban studies, but this question has largely been absent from the text itself. Here, I seek to specifically address the question of what might scholars of southern cities usefully draw on and what ought we reject. How can scholars distinguish between helpful and problematic northern ideas when seeking to undertake research both in and outside of southern cities? How can we make use of northern and other urban geographical insights without obscuring the very real differences between cities globally?

I have suggested that juxtaposition, unease, unlearning and learning anew is one useful frame through which to think about the process of developing theory from southern cities. In this chapter, I work through a parallel set of questions to help us better specify the precise component of theory that we might juxtapose, be uneasy with, unlearn and learn anew. I propose that unbundling the term "theory" and drawing attention to its component parts is one useful strategy for deepening engagements with the travels of theory. Postcolonial scholars, of course, are not alone in this search; one of the core tensions in social science remains diverging perspectives on what one can and ought to do with theory (Giddens, 1979; Tilly, 1984; Peet, 1998; Cresswell, 2012; Baronov, 2013). Scholars have long asked whether we can generalize across the world and find unifying principles as well as what can we learn from place-based ethnographic studies. Postcolonial (and decolonial) approaches more explicitly address the ways in which race, nationality and modernity shape the travels of theory, but I find it useful to view postcolonialism as relevant to, but not subsumed by, wider frameworks of enquiry.

Many scholars writing through the southern urban critique have articulated concerns with universalism. In this chapter I work with an unbundled version of universalism as well as distinguish universalism from totalizing theory. I use the term "totalizing theory" or discourse to describe what "usurps all other discourses (and all other belief systems). It presumes that its beliefs and assumptions about the world (and the model of the world that results from these) are the only correct ones" (Baronov, 2013: 174). In geography, modernist and Marxist approaches have been argued to provide totalizing accounts that serve as the foundational explanation which underpins all human activity (see Christopherson, 1989; Marden, 1992). I find it productive to use the term "universal" in a smaller way to describe what exists globally. This smaller term can be particularly useful when paired with unbundling the components of theory. In examining the specifics of arguments within the southern urban critique, some components of theory may be described as *universal* without necessarily claiming to be *totalizing*. Equally, we might be able to appreciate that some phenomena are actually universal: scholars may, for example, observe that the empirical phenomenon of inequality is universal; and who would argue against this? *Explaining* inequality, however, can be done either through totalizing accounts, or non-totalizing accounts open to causal explanations of inequality that differ across space and time. Even in a non-totalizing account, it is useful to investigate whether certain causes of inequality are or are not universally present. I believe that these distinctions are particularly useful for urban studies. Postcolonial studies originated in the humanities, but the very material conditions examined in urban studies make it important to specify when we are identifying the universality of objects, processes and patterns or the universality of explanations for them.

In this and the next two chapters, I work with the idea that unbundling the components of theory-making may well help scholars to better understand and explain to others not just whether theory can travel, but what specific aspects are better suited to mobility and how we might translate and transpose them. I start here with the literature on urban dispossession, displacement and gentrification as a point of entry into this conversation, a rather prolific area of scholarship in which scholars have been critically thinking about the travels of theory across the north and south. Urban dispossession and gentrification is not at the core of my area of expertise; I came to these readings as part of a wider exploration with my graduate students on the southern urban critique. My own inability to explain core points of difference prompted this deeper investigation into what exactly is being argued. In conversation with students I came to a reading that scholars have, at times, spoken past each other in this debate. This chapter is born out of these conversations and a more detailed analysis of select works with one graduate student: "we" in this chapter is Lené Le Roux and I.

My hope in working through my own struggles to interpret the precise arguments in this literature is to contribute to ongoing conversations seeking to better explicate and develop a middle ground between ideographic approaches and totalizing discourses (often at present being undertaken through the lens of comparison, see e.g., the special issue edited by McFarlane and Robinson, 2012). As is true throughout this book, my point in working out the arguments that follow is not criticism: these are difficult concepts and my insights are only possible because of the work of others before me. My hope is that offering an attempt at further clarity in the conversations examined below can contribute to more rigorous debate, dialogue and focused argumentation that enables us as a community of urban scholars to increasingly take on the challenge of incorporating southern cities more fully into urban studies.

I say "contribute" for two reasons. First, of course, it is too much for any chapter to unbundle all theories in urban studies. What follows is a selective analysis meant to demonstrate that terms are used differently across debates rather than a comprehensive analysis of dominant trends. Second, this is hard work, and often work that is not the heart of what we as scholars are most motivated to "get right." I include myself in this as high theory, epistemology and ontology are not where I am most confident and comfortable; this chapter has pushed me both to stretch and return the basics of how we put together arguments. This chapter is meant as a first effort rather than a final word.

Urban dispossession and displacement

Urban dispossession and displacement is one of the most prolific areas of study deploying the southern urban critique. As in many other fields of enquiry, scholars have sought to demonstrate the parochiality of northern theory through specific empirical cases. For example, Gillespie (2016) and Leitner and Sheppard (2018) assess the utility of "accumulation by dispossession," while Ghertner (2014, 2015), Bernt (2016) and López-Morales (2015) engage with the term "gentrification." As noted in the wider literature on the southern urban critique, many of these authors are concerned that urban scholars use northern theory too often and unreflexively: Leitner and Sheppard (2018: 437) call accumulation by dispossession the "default theory" used by urban scholars, and Ghertner (2014: 1558) describes gentrification as the "standard narrative" for explaining the globalization of land markets in northern and southern cases.

In working through this literature, we (Lené and I) first sought to separate when authors used specific terms to describe empirical phenomena and causal theory. Gillespie (2016), for example, clearly sees an *empirical thing* that fits the widely described phenomenon of accumulation by dispossession. But the *causal theory* of accumulation by dispossession, he argues, does not fit. Instead, "accumulation by urban dispossession in Accra serves a fundamentally different function to the 'classic form' of primitive accumulation described by Marx" Gillespie (2016: 74). In other words, we can see the empirical thing happening, but its motives, causes and effects are not the same as what has already been articulated in the literature based on studies of northern cities. In Gillespie, therefore, we see an argument for retaining the term "accumulation by dispossession" as an empirical phenomenon but widening the range of possible scholarly explanations for why and how this occurs.

We similarly see scholars grappling with how to articulate the difference between empirics and explanation in a related debate on gentrification in southern cities. We identified a plethora of categorical descriptors for the term gentrification: it is a concept, a framework, an explanatory device, a narrative and so on. López-Morales (2015: 565), for example, explains that gentrification is an explanatory device, "a vivid and mutating process and set of accompanying theor-izations" and both he and Ghertner see it as a "lens" through which urban change or phenomena are analyzed. Bernt (2016) is most explicit about the challenge of these multiple uses: he argues that we need to see gentrification as an empirical phenomenon, and then provide multiple explanations for why this occurs based on the study of cities globally, an argument similar in structure to Gillespie.

Separating the empirical phenomena from causal theory enables us to better understand the precise critiques present in these articles. They agree that a thing widely understood as the movement of urban residents (displacement or dispos-session) is happening and that it looks on the surface similar to patterns observed in northern cities. While López-Morales (2015) may be understood to offer a classically Marxist analysis in which all other factors are subsumed under the influence of capital, others find such a totalizing narrative insufficient. What binds these latter scholars together is that they are concerned with the causal explanation implicit in northern theory: even if the empirical phenomenon looks similar, their cases suggest that a much wider range of explanations is needed to adequately explain their southern examples.

What, then, ought we do? The answers drawn from the literature reviewed above differ; they deploy different tactics for recognizing southern difference. While Gillespie suggests modification of a term to include different causal explanations for a phenomenon, Leitner and Sheppard urge extension and new terminology, and Ghertner argues against deploying northern language. For Gillespie, provincializing Marxist concepts ought not lead to their rejection; he proposes instead a modification as existing theories are useful but partial. In their abstract, Leitner and Sheppard (2018: 437) frame their intervention as "an extension," but offer alternative vocabulary to emphasize this distinction: rather than accumulation by dispossession, "contested accumulation through displace-ment," they argue, "is better attuned to capture the distinct features of Southern cities." Ghertner (2014), in contrast, argues against extending gentrification to capture wide-ranging explanations (see also Lees *et al.*, 2016). Doing "so sheers gentrification of its analytical specificity that it loses both its explanatory power and its political potency" (ibid: 1554). This is not to say gentrification ought to be discarded on the whole, but instead, retain its analytical specificity (see also my arguments above about the city and modernity). Our point here is not to criticize these authors: again, this is difficult work. There is limited space in any article and the authors have done much heavy lifting in the pieces reviewed.

Ghertner (2014) also reminds us that our enquiry is not only an academic exer-cise. The narrative of gentrification (his focal point, although this point is true for accumulation by dispossession as well) is not only an analysis of what has happened and why, linking capital accumulation with particular state relation-ships. Embedded within this narrative are also *political responses* that seek to redress gentrification. Bernt (2016: 638) explains that advocates of gentrification theory see it as "an indispensable armoury for struggles against urban injustices experienced at a global scale." We may then see the causal explanations as includ-ing not only a narrative that explains what is happening, but prescriptions as to what ought to happen to achieve a more just city in the context where capital-ism is seen as the dominant force creating and limiting urban change. Getting the analysis right matters, then, because it shapes the framework for action. Build-ing on this, we might also attend to the concern that northern-derived critical frameworks often *also* include northern-derived answers for what to do, including prescriptions identifying what the normative goals of action ought to be. I return to this point in Chapter 9.

In seeking to abstract from the literature on dispossession and gentrification into wider generalizations, then, we see that Ghertner, Gillespie and Leitner and Sheppard make clear that scholars ought not a priori assume that causal theory derived from northern cases is explanatory. Instead, scholars ought ask whether, whether and what a theory usefully explains, as well as where it is useful. Beyond this, scholars ought to also attempt to explain what lies outside this framework. The tactics deployed by these scholars, however, differ in terms of whether spe-cific terms ought to be rejected or modified. This concern surely resonates with general scholarly questions as to when to repurpose or create a neologism (see also Chattopadhyay, 2012), but as argued throughout this book, such questions must be read through a postcolonial concern. Creating different terms might well serve to reify difference (this is a southern theory, not applicable to other cities). It may, however, be equally difficult to alter an already-existing term to wider usages. Both of these are imperfect tactics, and here is not the place to provide a general rule; instead, my concern is that we would benefit from more explicit con-sideration to the question of incorporation, expansion or neologism within and across different literatures.

More broadly, this chapter has sought to highlight three components of the-ory that are at times bundled together and/or conflated across ongoing dialogues. There are empirical phenomena: these might be objects (houses), or processes (the sale and purchasing of homes). There are causal narratives: these tell us *why* a certain house is where it is and/or why it was sold and purchased. Finally, there are normative judgments: these are at times developed from the causal narrative, but also include consideration of sociopolitical context and particular moral judg-ments about a just and desirable world. More careful identification of which of these is the subject of scholarly attention might well be useful in clarifying points of convergence and divergence in urban scholarship.

In the next chapters, I work through four of my own scholarly interventions with this wider framework of unbundling and specifying the components of theory in mind. I seek to show how my own thoughts about environmental justice, urban appropriation, infrastructure and the nexus of labour and income have changed as I thought through southern cities. I both seek to develop a deeper understanding of how I came to particular ideas as well as clarify whether the point of argument-ation is about empirical difference, causal explanation or normative judgement. My intent is to demonstrate a useful line of investigation for urban studies beyond these particular topics and contribute to clarifying ongoing processes of urban theory-making.

8 Theorizing anew through southern cities

In this chapter and the next, I work through several of my own efforts to think about the travels of theory through an investigation of four rather unrelated themes in urban studies. Three of these are widely used terms in the literature: environmental justice, urban appropriation and the modern infrastructure ideal. The fourth theme is more difficult to label. It is an idea that underpins, but is rarely a focal point, urban investigations: the relationship between work, virtue and the politics of distribution.

I have written more detailed arguments thinking through each of these concepts from a southern perspective and case, and my purpose here is not to rehash the details of published scholarship. Instead, I focus on articulating the process of juxtaposition, unease, unlearning and learning anew as well as the components of critique and line of argument developed in each case. I seek to draw out specific arguments about what might beneficially travel and what might be provincialized, and how this enables new insights into what cities are, how they operate and what residents aspire for them to become. This means, at times, reading beyond the published texts, and hence the ongoing egocentric approach which I hope the reader will continue to forgive! This chapter focuses on the first two concepts and my arguments here which are more conventionally "theoretical" (i.e., are focused on generalizable explanations of what we see). The second two themes, addressed in Chapter 9, examine the relationship between theory and practice through an investigation of the normative judgments that often (intentionally or not) underpin and travel with theory.

Environmental justice

My first foray into the limitations of travelling theory focused on the use of the term "environmental justice" in South Africa. This work is largely rooted in my own experiences with the NGO Earthlife Africa (ELA), beginning in 2002 and intermittently until 2013. Environmental justice as a phrase has been used to describe social movements and scholarly analysis internationally and globally (Bullard, 1990; Cock and Fig, 2000; Schlosberg, 2004; Pellow, 2007; Myers, 2008; Ekhator, 2014). It is a phrase explicitly formulated in the north and written about first by scholars in the north and has travelled widely to and across the global south. Embedded in the concept of environmental justice is a description of an empirical phenomenon, a correlation between environmental benefits, burdens and demographics. Activists and scholars have equally sought to explain this phenomenon, most typically through examinations of class and structural racism. Implicit in the use of an environmental justice frame, then, is an explanation of a particular spatial pattern. No doubt the empirical phenomenon of disproportionate impact and explanations of it, however, has an equally long southern history (e.g., Crosby, 1986; Carruthers, 1995; Grove, 1996; Dovers *et al.*, 2003; Jacobs, 2003); I return below to questions of the multiplicity of explanations for "environmental justice" through this wider lens. My interest here differs from those arguing that the phenomenon is widely demonstrable: instead I focus on tracing what moved with the term as it flowed through an examination of "environmental justice" in scholarship on South Africa.

Might there be problems with the travels of environmental justice? Does it carry totalizing or universalizing tendencies – and if so, is this problematic? Is there unlearning to be done when transposing environmental justice analysis onto southern cases? What, if anything, should be provincialized? I am not the only one to have asked such questions, and many who have used the term have thoughtfully observed ways in which it stretches as it has moved (see Schlosberg, 2004; Myers, 2008; Walker, 2009). Here, I discuss my concerns with its travels as a way to more carefully think through which components of the bundle "environmental justice" might benefit from provincializing. I draw on Lawhon (2013a); like many others, in the original piece I bundled and conflate theory, concept, social movement, explanation, connotation and politics. In this chapter, I seek to work through a more precise articulation of my analysis with a focus on what an analysis of environmental justice in South African scholarship shows us about the travels of theory. Specifically, I work to show how we might better account for the benefits and limitations of travels through unbundling the components of theory.

My volunteer work with ELA provided my first deep engagements with environmental justice. I was twenty-one and had encountered the term during my undergraduate studies, but only in a peripheral way. ELA self-identifies as an environmental justice organization, a point that my mentor Muna Lakhani used to differentiate ELA from other "green" organizations. For ELA, environmental justice was a social movement category, a label through which to identify allies. It also shaped which to view environmental issues: for example, while there are many concerns that can be elaborated about nuclear power and genetically modified organisms, ELA focused on the disproportionate impacts (broadly writ) on communities of colour.

My experiences with ELA were formative to my understanding of environmental justice, and these experiences constituted the basis from which I encountered the scholarly literature. This was no quick process: a decade passed before I attempted to articulate my intellectual struggles through this juxtaposition and its limits. It was prompted by my own efforts to learn more from the scholarly literature on the wider struggles of ELA, and a disappointment that its history was not particularly well-documented in academic discussions (although see Hallowes, 1993; Steyn and Wessels, 2000; McDonald, 2004).

When I searched, a decade or so ago, for environmental justice cases in South Africa in the international academic literature, I did find many examples. As I demonstrate in a more detailed review of the literature (Lawhon, 2013a), environmental justice scholarship tends to begin with environmental justice theory and show the ways in which the South African cases exemplify or are exception to established explanations. The work has encouraged productive reframings to make environmental justice more relevant to South Africa, and suggests that environmental justice as an analytical frame has relevance. Yet, much of the work conveys a sense of dissatisfaction, a sense that there are things as yet unexplained. That sense very much resonated with my own disappointment that the struggles I knew best were under-accounted for in the scholarly literature.

One case stands out in the South African environmental justice literature. South Durban is undoubtedly the most commonly researched environmental justice topic, and possibly the most common struggle written about in South African environmental social science (cf Scott *et al.*, 2002; Barnett, 2003; Barnett and Scott, 2007; Scott and Barnett, 2009; Leonard and Pelling, 2010; Aylett, 2010a, 2010b). I argued in 2013 (Lawhon 2013a: 132), "Activists and scholars consistently employ the language of EJ as the key frame to describe the South Durban controversy over the citing of oil refineries, chemical manufacturers, toxic landfills and chemicals storage facilities near a 'community of colour'." Why this issue has garnered so much attention is less evident. It is clearly not the most egregious environmental injustice if compared to forced removals from protected areas (Carruthers, 1995), the environmental health impacts of asbestos (Braun and Kisting, 2006) or informal settlement flooding (Dixon and Ramutsindela, 2006). Nor it is widely representative of environmental injustices in South Africa as those disproportionately harmed are largely middle class and a racial minority (Indian). While impossible to clearly identify the reason why this case has garnered so much attention, I suggest that the salience of this case can likely be attributed to the fact that activists have made their controversy legible to Northern activists and scholars. While the EJ framing clearly draws the attention of researchers in and outside South Africa, other examples of complex environmental issues – which may have greater social, economic and environmental impacts but fewer links to well-articulated framings and literatures – have attracted much less attention."

In the paper, I point to two primary concerns with this observation. The first is political: framing a problem as "environmental justice" probably resonates better with international environmentalists than the average South African. In South Africa, "the environment" continues to connote conservation concerns with protecting charismatic megafauna and is often read as a middle-class, white concern. I believe that this trend is less true today than a decade or two or three ago. However, my own various works on environmentalism in the media, as well as personal experiences with fellow activists and teaching environmental topics in South Africa, suggest that there is still reason for concern (see Lawhon and Fincham, 2006; Lawhon and Makina, 2017; Lawhon *et al.*, 2018a). In the 2013 paper

(Lawhon, 2013a), I argued, "If Holifield (2009: 652) is right that the test of political relevance and usefulness of a theory is 'whether it is taken up by those it has sought to represent, and whether it enters into the ongoing composition of society,' then we must critically reflect on whether framing environmentalism in line with international discourses may actually alienate people from broader livelihoods struggles." Notably this is not a question about veracity, but relevance and usefulness.

More substantively for the argument of this chapter, I also examined the implications of the dominance of this case for the construction of environmental justice explanation. I argued in 2013 (Lawhon, 2013a), the pattern appears to be instead of searching for a broadly construed environmental justice case from which to learn, scholars look to what is most familiar (in terms of the literature), which might well also be what is most easily publishable through an established intellectual genealogy. Again, not all South African environmental justice scholarship has studied this single case. For example, Dixon and Ramutsindela (2006: 134) have explicitly argued for the importance of casting a wider net when seeking environmental justice cases as well as a focus on just how difficult it is to create "environmentally just" outcomes. They complicate any easy narrative of redressment through a compelling analysis of "whether environmental justice would be achieved by upgrading an informal settlement in a flood-prone area or by resettling residents away from the flood area." Their argument is a productive example from which to think about how the travels of environmental justice might be troubled through cases that do not fit expected patterns.

In sum, this initial effort at thinking about the travels of theory was sparked by my own experiences working with an environmental NGO that self-identified as an environmental justice organization, but did not primarily work on environmental justice issues that resemble the canonical global north cases. My argument is not that scholars were wrong in their use of environmental justice to describe an empirical phenomenon or explain it. It is that if the selection of cases – of the empirical phenomena – by scholars implicitly focuses on those that fit our pre-existing gaze, we probably miss cases that might provide new insights into environmental justice. Such new insights might include, for example, different explanations of causality or different perspectives and tactics through which to redress injustice.

My argument about the travels of theory here point to, rather than addresses, a research gap. To some extent, my work on electronic waste and environmental justice builds on the ideas developed above (see Lawhon, 2013b). Environmental justice advocates had brought the issue of global e-waste flows into northern media discourse (Iles, 2004), but my work complemented others who complicated this narrative with questions about reuse, the digital divide and economic development, and the extent to which ecological modernization and environmental justice might be seen as less antagonistic (see Lepawsky and McNabb, 2010; Wang *et al.*, 2012; Lawhon, 2013b). In the next section, I focus on work that is more explicit about showing what happens when we start from unexpected empirics, empirics that are not readily analyzed through existing causal explanations. This work addressed a very different term (urban appropriation) but follows a similar line of argumentation: I point to the importance of capturing what falls outside existing explanatory frames by looking at different empirics. In both cases, the *empirical phenomenon* of environmental injustice and urban appropriation appears to be present across the north and south. However, in both cases, it is less clear that existing causal explanations are sufficient to capture the diversity of ongoing processes (what Leitner and Sheppard (2018: 451) delightfully describe as "insufficiently capacious"). Unlike in the 2013 paper (Lawhon, 2013a), however, I then take the next step and work to use these different empirics as a starting point from which to develop a new modifier to emphasize this different explanation.

Agonistically transgressive urban appropriation

My work on urban appropriation is rooted in my daily walks in Tshwane between my flat and the University of Pretoria. Intersections with "robots" (traffic lights) in South African cities, as in many other parts of the world, are often busy places where people seek donations from those in the cars passing by. At the corner of Eastwood and Pretorius Streets, near a small shopping complex with a Woolworths, a nice cafe and a high-end souvenir shop, an older white man was often the only beggar in sight. He received more handouts than any such person I had ever seen: there were often more windows cracked and hands with coins than he could collect before the relevant robots changed to green. Why, I wondered, was he alone so often here? How did he hold such prime "real estate," or, I learned to ask, appropriate so much urban space? I never got the answer to this question. One day he was gone, and the intersection became populated by several younger, black African men.

The question, however, did not disappear. Instead it morphed into a wider enquiry into the appropriation of space in South African cities. This includes work with PhD candidate Anesu Makina and our wider research project, "Turning Livelihoods to Rubbish" (with Erik Swyngedouw and Henrik Ernstson) as well as conversations, again, with Joe, a more classically trained urbanist who had also written on rights and permissions in urban space.

Joe, Anesu and I talked through our concern that what we saw in South Africa was inadequately described through existing frameworks. That this was an empirical phenomenon of appropriation seemed straightforward: claims were being made to an intersection, and clearly some people were permitted and others were not. This was not a legal right, but a possession in practice. The appropriation appeared to me, as a regular passer-by, to be respected by others as far as I could tell, but surely this was not a permanent state of affairs. Translating this to our waste project, informal waste pickers in city streets made claims to rubbish bins and their contents. In both cases, there were surely more people seeking access to such spaces (and the associated resources) than there were spaces available. And yet, conflicts were typically not visible to a casual observer.

Conceptualizing the possession of space as durable rights, or rights to strive for, did not seem to capture the ongoing dynamics. Lefebvre notions of urban appropriation seemed to expect too much regularity, consensus and a more present, active and consistent regulatory state. Neither, however, did Bayat's idea of quiet encroachment seem to apply, even though this concept was developed from global south cases. Seeking donations at intersections and waste picking are quite visible, organized, practiced by a fairly consistent set of persons and, we assumed, regulated by a set of social codes that were unclear to the casual observer. Further, the process of regulation seemed to largely take place beyond the state, between at least some marginalized urban actors (although possibly connected to powerful ones; see Seekings, 2001; Shearing, 2001; Drivdal and Lawhon, 2014). The rules developed were probably not static, not written and probably not easily shared with an outsider (I doubted I would have gotten a direct and accurate answer at my neighbourhood intersection, especially if an actor with more physical power was taking a cut of his gains, without extensive relationship building first).

Joe, Anesu and I therefore developed the concept "agonistically transgressive urban appropriation" to fill in this gap (there are some interesting parallels here with Leitner and Sheppard's "contested accumulations through displacement" reviewed in Chapter 7, which appears to have been written at a similar time). We retained the term urban appropriation to suggest that a similar phenomenon was happening: appropriation is probably a universal phenomenon in cities. And yet, how it operated and was regulated was not explained by the existing literature: this was not a disagreement over rights or an opportunistic taking. Lawhon *et al.* (2018b) argue that it is not that other modes of appropriation are always and entirely absent from and irrelevant to South African cities: one might find such examples, and the literature suggests that there might be utility in doing so. We were concerned that *starting* with cases that exemplify pre-existing explanatory frames leaves much unexplored. Recognizing different modes of appropriation requires choosing cases that fit less neatly and being open to alternative explanations of how and why appropriation occurs. It also requires being open to different explanations for how and why approprations are contested.

The analysis here has parallels with the literature on dispossession covered in Chapter 7 and that of environmental justice above: even when the empirical process of appropriation looks similar, the explanation differs from that provided by established theory. We therefore argued for retaining the term "urban appropriation" (to describe an empirical process) and are open to the possibility that both Lefebvre's articulation and Bayat's articulation might well apply to other southern cases (i.e., be present across many different kinds of spaces). We argued, however, that neither Lefebvre nor Bayat provided totalizing explanations nor were universally *explanatory* (i.e., not explaining every case of appropriation): some urban appropriations are not explained by either existing framework.

This work, then, advances on what I proposed about environmental justice several years earlier: instead of merely calling for more diverse case selection and alternative explanations, it develops a new explanation through the purposeful selection of cases that do not easily map onto existing explanations. Of course, identifying this "gap in the literature" came from being in the city, not seeing the city through existing literatures (see Pierce and Lawhon, 2015). Admittedly, the selection of different empirics made them harder to think about and harder to write about. We will see how difficult it is for Anesu to publish her empirical work (it is worth noting that her PhD committee has thus far been supportive of our exploration, for which we are grateful!), and I return to pragmatic questions about training and writing in the conclusion. In the next section, I work to think more explicitly about differently structured arguments: I seek to more deeply engage with the ways in which certain norms become unexpected passengers when theory travels.

9 Recognizing and reworking the norms that travel with theory

Thinking about the travels of environmental justice and urban appropriations, as described in the last chapter, has usefully helped me to deepen my analysis of scholarly explanation. But, at the end of the day, I am most interested in the tangible implications of my work. I want to get the explanations right not because I want to understand the world better, but because I think we collectively can build a better, more just world (see Castree *et al.*, 2010). No doubt this motivation bears the weight of a fraught, racialized history: there is no shortage of evidence of colonial actors working in the belief that they too were building a better world, and there is no shortage of racism and paternalism underpinning much contemporary development practice. Many of my co-authors and other colleagues have noted their unease with analysis that enquires as to what works better. No doubt the concerns about developmentalism from critical development studies have usefully carried over into southern urban studies!

And yet, the realization that change has been problematically instigated in and through the academy in the past and at present is a reason for deepening our investigations of scholarly practice, not for disconnecting the academy from practice (see McEwan, 2003; Kapoor, 2004; Nagar, 2014). At the very least, the co-authors and colleagues with whom I have discussed such concerns are equally concerned that simply elaborating what is there, without contemplation of what might make for a better world, is a thoroughly extractive academic process. Drawing norms and practices into the realm of our scholarly praxis raises important questions about how to identify and articulate what matters and what we care about when trying to work towards change. Plenty of ongoing work on participation is useful for helping us understand better practices here, and my hope is that building on these concerns to more clearly separate empirical phenomena, causal explanations and normative judgments will usefully contribute to such conversations. Refusing to engage with questions about what to do, however, is neither just nor what most southern scholars and research participants I have encountered want.

Various scholars contributing to the development of the southern urban critique have shown the importance of getting the explanations right, as misdiagnosing the causes of problems probably means developing ineffective responses. Ghertner's (2014) writing on gentrification theory, for example, points to this connection:

he argues that we need to attend more carefully to the travels of theory in part because strategies to oppose gentrification in cities in the global north are unlikely to succeed if the causes of gentrification are different. Relatedly, my investigations of environmental justice reviewed in Chapter 8 also raise tactical questions: even if the analysis of a problem travels adequately, the language used in framing the problem may not translate well.

In this chapter, I build on these concerns with practice, but from a different angle. I seek to interrogate the "better world" that is often implicit in different theories and that often travels with them. I draw inspiration here from the work of Chattopadhyay (2012: xx), whose *Unlearning the City* seeks, in part, to "demonstrate the particularity and situatedness of the dominant frameworks that are passed off as global norms and criteria of desirable outcomes." Much of the concern from scholars of southern cities with the world cities literature is, surely, concerned with the accuracy of analyses; my sense is that scholars are eqaully concerned with the barely concealed judgments of northern/world cities as good cities (Robinson, 2002; McDonald, 2008). Many scholars have begun their writings about the southern urban critique with an observation that southern cities are judged for their failings. At times, this has meant something like "southern cities are excluded from theory because they are deemed to be failures," yet this is largely nonsensical, as northern "failings" have not been excluded from contributing to urban theory. Instead, I understand this frustration as a gesture linking theory to practice: southern scholars are tired of being told how their cities ought to be based on what has happened in northern cities. To some, this point might be incredibly obvious. Yet, I think many scholars who work primarily in the global north (and often feel that their work is rather dissociated from what happens in cities) miss this concern. For scholars and residents of southern cities, the experience of being told what cities are and ought to be, based on northern referents, remains powerful.

I have struggled some with my word choice here: Graham and Marvin (2001), for example, use the word "ideal" to describe the vision of universal, uniform, networked infrastructure. Of interest here is the transformation of the ideal they describe in the north into a norm in southern contexts. My choice of the term "norm" is meant to emphasize social expectation and its association with assumptions about how things ought to be, similar to Chattopadhyay's usage above. I also adopt this term for its connotation with normal (typical plus socially appropriate), and the provocation that comes from thinking about the spatial travels of the frame of reference for norms. The norms I speak of here are not a product of what is typical ("normal") in the global south, but a translation of what is presumed typical and appropriate in the north into aspiration for the south.

Here, then, I examine the often-implicit links between what we value, what we question and what we let stand, and how these choices shape our political imaginaries. This is not to suggest that there are not relevant questions about the norms embedded in theories of environmental justice and urban appropriation; it is to say that my published work thus far has not sought to work through them. I believe there is utility in deeper examination of both explanation and norms. Therefore, in Chapter 8 and in this chapter, I demonstrate the utility of contributions that can be developed through both lines of argument rather than an evaluation of better and worse lines of argument.

While in Chapter 8 I was able to come to different explanations through unlearning established explanations and seeking to explain what was not readily explained, identifying and provincializing norms is an even more fraught process. Again, it is through juxtaposition that we might begin to see what is so deeply embedded that we cannot see it ourselves without help. Juxtaposition alone is insufficient; again, it might well result in recognizing and embracing rather than unlearning norms. Here I seek to work through sitting with unease as part of being able to work towards unlearning and learning anew.

I start with an examination of the modern infrastructure ideal, identified in the literature as a prescription of how infrastructure ought to appear and operate. Critical urban infrastructure scholarship has articulated and examined the travels of this ideal across the north and south. In the section that follows, I examine the relationship between work, virtue and the politics of distribution. I work through my own surprise, discovery and a sense of new imaginaries that came from disentangling, historicizing and provincializing this relationship. My intention is to use examples from my own work as a lens through which to begin unbundling theory and clarifying the arguments that have shaped the southern urban critique. My hope is that this enables similar enquiries in other areas of study.

The modern infrastructure ideal

Infrastructure studies are a particularly productive arena in which to investigate the normativity embedded in theory because an ideal, and its travels, has already been identified and investigated. Graham and Marvin (2001) consider the modern infrastructure ideal to be the global social and political goal of achieving universal, uniform, networked access to services. Since this important articulation of this ideal and its "splintering," scholars of southern cities have sought to add more nuance to our understanding of the travels of the modern infrastructure ideal. They have called attention to the hegemony of this ideal as discourse and policy goal (Jaglin, 2014; Furlong and Kooy, 2017; Anand *et al.*, 2018; Lawhon *et al.*, 2018c) as well as analyzed its travels and associated transpositions (Bakker, 2003; Boland, 2007; Jaglin, 2008; Nilsson and Nyanchaga, 2008; Monstadt and Schramm, 2017).

Many scholars writing about southern urban infrastructure have critiqued the modern infrastructure ideal in ways that parallel the structure of arguments in Chapter 8. For example, Furlong and Kooy (2017: 891) argue that a scholarly focus on networked infrastructure has meant that we have given little attention to explaining "the many ways that water is accessed, urbanized and fragmented." In other words, looking primarily at the empirical phenomenon of formal/modern infrastructure, and its lack, limits the possibilities for scholarly enquiry and explanation. Their central concern is that over-emphasis of modern infrastructure and its failings obscures other urban practices and efforts to understand the predominant modes through which urban residents access water. It also means we miss many forms of environmental inequalities: for example, instead of simply access or not, understanding water to flow through multiple modes of infrastructure raises questions about the varying qualities of the urban water (see Tiwale *et al.*, 2018).

This line of argument parallels those outlined above for environmental justice and urban appropriation: if, in studying southern cities, we only look at what is already explained in northern literature, we miss much of what is happening in the south. And empirical differences – things that are not easily captured by existing explanations – hold the potential for new analytical insights. In response to this concern, various scholars have called for increased attention to, and written about, everyday practices in order to better understand the ongoing realities of service access and provision in the global south (see Graham and McFarlane, 2014; Lawhon *et al.*, 2014; McFarlane *et al.*, 2014; Silver, 2014; Truelove, 2016; Alda-Vidal *et al.*, 2018; Baptista, 2019; Pihljak *et al.*, 2019).

It is clear that this work responds to an analytical gap: focusing on how to get people connected to modern networks (whether through the state or private sector) does not tell us much about the realities of life for most residents of southern cities. Working through how people actually access services helps us to understand the processes and practices that shape everyday life. This literature is less explicit about its normative ambitions, however. When it comes to infrastructure, many of us have been cautious about questioning whether the modern infrastructure ideal is actually a good goal. No doubt uniform, universal, networked infrastructure is not the typical mode of access to services for most residents of southern cities. But does this mean we ought to reject the ideal of state-supported uniform, universal, networked services? If so, why? Is this because it is unrealistic, or because it is undesirable? And then: is there a new normative ambition – something beyond the modern infrastructural ideal – that might replace the old one and help shape new imaginations of urban infrastructure?

My own recent work with colleagues (Lawhon *et al.*, 2018c) seeks to push us more deeply towards these questions. My coauthors and I explicitly urge us to collectively question the assumption that safe and reliable services are best provided through a particular infrastructural configuration (uniform, universal, networked). I say emphatically that the purpose of the article is to frame a line of enquiry, not presume that we know the answer, let alone describe this alternative. We argue instead that as scholars we ought not only attend to the empirics on the ground or the existing ideas (and their rise, fall, and travels). Without presaging conclusions, we ought to be open to the possibility of, and willing to investigate, other modes of infrastructural provision which may – *may* – have more just and sustainable outcomes. These may be already existing modes, but we may also learn from existing practices in order to develop ideas of how things might work better.

We offer heterogeneous infrastructure configurations as both an analytic and, potentially, a way of imagining new alternatives. In the first sense, we encourage others to see urban infrastructure not as a single, located site, but as a range of options with differential availability, costs and benefits (a configuration, even when not a network or grid). In my own home in Oklahoma, we get energy from electricity, gas and batteries, and this means that if electricity is disrupted, we can still cook (with gas), use flashlights (with batteries) and be entertained by dancing plastic robots (the latter of which is, surely, less significant…). When I discussed this work with students at the University of Miami, they (laughingly and happily) described their choices about which bathrooms to use on campus, noting that they often walked out of their way to preferred sanitation options. In global south contexts, infrastructure is even more heterogeneous in type and quality, and disruption is omnipresent, an expected part of most infrastructure (spurred by technical, ecological, financial and social dynamics). It is common to have multiple sources and sites for sanitation, energy, water and so on (which is not to say any of them are "ideal"). We argue for understanding the accessing of infrastructure as a set of options that residents navigate, recognizing that choices are shaped by myriad types of constraints.

Lawhon *et al.* (2018c) follow this argument with the suggestion that framing infrastructure as heterogeneous infrastructure configurations opens our imaginations to new modes of understanding and enabling access to infrastructure. Rather than seeking the perfect toilet or the single reliable and affordable energy source, or even the single best social configuration for ownership and payment, might we imagine configurations that enable better access under a diversity of always-changing, always-constrained circumstances? Our point here is not to suggest that everyday practices already happening in southern cities are necessarily better, but instead to shift from description and analysis of how infrastructure is accessed to thinking about what might enable more just, sustainable, reliable and resilient heterogeneous infrastructure configurations in the world we have.

How might scholars try to understand whether heterogeneous infrastructure configurations might be more just and sustainable than the modern infrastructure ideal? Most importantly for the argument in this book, it first entails a precise articulation of the normative ambition. This starts with separating the mode of achieving a goal from the goal itself. In other words, I believe the collective goal is not uniform and networked infrastructure; this is a modernist ideal, and one we might displace. We might reject the modern infrastructure ideal as a mode of achieving services without rejecting the ambition of achieving safe, reliable, universal services. It is worth restating that as authors of the paper, we are far from advocating a new mode; we are more confident about arguing for such explorations. We also find inspiration in connecting our insights to work in the global north seeking, for example, energy mixes and water recycling and infrastructures that work with, rather than seek to control, nature (e.g., Nilsson, 2019).

More broadly, my intention in this chapter is to argue that the southern urban critique includes not only questioning explanations, but also identifying and questioning the norms that underpin our assumptions of what cities ought to be. These norms are not always explicit, but sometimes travel in hidden, unintended ways. The modern infrastructure ideal links the questions raised in Chapters 3 and 4 about modernity with the idea of unbundling our arguments, and it is useful to return briefly here to this intervention. I have argued that retaining a narrow use of the term modernity to describe a particular set of norms and practices is analytically important, for it enables us to label a particular (if fuzzy at the edges) mode of social processes. As a totalizing discourse, modernity provides an inaccurate and problematic gaze on the world. It is flawed both as an explanation (its articulation of the way the world is and our ability to control and predict it) and as a normative judgment (in its moral perspective about the hierarchy of who matters and why). It is particularly important that we disentangle these components. I am arguing for retaining *modernity as a specific analytic* without conflating this with a *judgment that modernity is accurate or good*. Doing so helps us to take pieces of modernity forward into new modes of being and seeing the world: we might retain its belief in the social utility of technology while rejecting the idea we can control nature; we might retain its belief in universal justice while rejecting the idea that this means the same thing to everyone. We might reject the idea that the city is modern, and the hierarchy that makes some cities better than others, without either rejecting the city as an analytic or attempts to articulate its (fuzzy, dynamic) borders, processes, and how and for whom it works. We might reject the modern infrastructure ideal, without rejecting the importance of material flows in enabling dignified lives for urban residents.

Here, I have worked with the idea that the southern urban critique is not only about the production of theory as conventionally articulated in the academy. It includes a critical examination of the norms that have long travelled embedded in explanations of what has happened elsewhere, and a more detailed understanding of why they might be inappropriate. Such critical examinations include thoughtful analysis of whether southern cities *cannot* mimic northern ones or *should not* strive to be like northern cities, or plausibly both. My argument here is that we ought not reject the overarching idea that scholarly enquiry might contribute to displacing norms and practices as well as the development of new norms and practices. Unlearning and learning anew can help us to better navigate the murky waters identified in critical development studies in terms of how we might learn across difference. More careful attention to what is implicit in "theory" and its unexpected passengers can help us to provincialize colonial modern norms and open spaces for articulating and envisioning alternatives that might better resonate with the residents and realities of southern cities – and possibly, northern ones too.

Work, virtue and the politics of distribution

Critical infrastructure studies in the global south have been explicit about the norm that is under investigation: the modern infrastructure ideal of universal, uniform, networked infrastructure. Yet, as I noted briefly above, the norms that shape our scholarship are often much more difficult to see. Here I sketch a recent set of questions that has unexpectedly arisen from my own research on urban waste, waste politics and livelihoods. Through unexpected juxtapositions and much unease, I have come to a set of ideas that sprawl far beyond discard studies, beyond the global south and surprisingly include the 2020 United States presidential campaign. But let me first take a step back.

I started my research on waste in the global south with a political stance similar to many others: I thought drawing attention to the positive impacts of informal waste collection might contribute to increased bargaining power and improve workers' ability to claim more resources (see Millington and Lawhon's, 2019 review of waste studies). Yet my work on electronic waste recycling, briefly noted in Chapter 8, raised questions that made for a more complicated puzzle. One of the key issues that emerged from my dissertation data, although I struggled to incorporate it into my analysis, was the relationship between technology and employment. Many well-intentioned owners of e-waste small businesses observed that recycling e-waste with more sophisticated technology tended to reduce ecological harm and health impacts on workers and increase recovery rates. But investments in technology typically meant that fewer people were employed in the industry. As but one example, a small business owner reported that he had tried to hire workers to strip wires manually, but this simply could not be done at a profit in his case. This was not an instance of a greedy capitalist business owner: he already knew that profits would be maximized through investing in technology, not labour. His hope was that he might sacrifice some profit for increased employment, although in this case the math did not work out in this way (see Lawhon *et al.*, 2019 for similar stories from our recent waste work; see also Robbins, In press on technology, labour and political ecology). During this time I also became aware of a project in Cape Town, where I was living, piloting a two-bag household waste collection system to be run by a private waste company (i.e., recyclables in one bag, waste in another). It was clear that doing so would displace many informal waste workers (many of whom were intended to be contracted to sort recyclables) but also probably increase municipal recycling rates (see Tischler, 2013).

These questions formed the basis for a new research project with colleagues from the University of Cape Town and the University of Manchester, "Turning Livelihoods to Rubbish?" Our title was intentionally ambivalent, questioning whether more people should be turning towards waste as a source of income as well as whether politics and technology are displacing existing livelihoods. Early conversations with the project team brought our different perspectives to the forefront of the project. In contrast with most scholarship on informal waste work in the global south (see Millington and Lawhon, 2019), some members of the project team argued that informal waste picking was not a job that we should seek to protect.

As with the work on infrastructure above, I sought to separate the mode of delivery from the underlying objectives. Did we agree that recycling should be pursued? Sure, albeit mindful of the limits of recycling. Did we agree on livelihoods for everyone? Again, yes. Some of the members held more classically Marxist beliefs and pointed towards the importance of better organizing and more formal jobs. My own long-term work in South Africa made me deeply skeptical of such a position: for decades, the state and international development agencies have been working to redress high unemployment, but with limited success (although public and scholarly discourse continues to be largely directed towards growth and employment, e.g., Seekings and Nattrass, 2019). While individual projects might succeed, on the whole, the employment situation is worsening (Banerjee *et al.*, 2008; Anand *et al.*, 2016). From this perspective, bad work is better than no work.

The juxtaposition that sparked my unlearning here was two different perspectives: one rooted in long-standing northern scholarship and politics and the other in a resigned acceptance of ongoing southern employment realities. Neither of which was particularly satisfactory. Neither seemed likely to actually produce the outcomes we agreed were good: dignified livelihoods and materials recovery. My own unease was rooted both in dissatisfaction with the options on the table and differing opinions amongst the research team. My searchings for reconciliation, then, had my team members in mind as an audience, and I am doubtful I would have started on this irregular search if our project team all had a united position, even if it was an unsatisfactory one.

Joe Pierce enters the story again here. He pointed me towards a wider literature on automation (or more specifically, technological disemployment, the empirical phenomenon of mechanical labour displacing human labour) in the global north, and shortly afterwards, a series of books were published providing useful overviews (Brynjolfsson and McAfee, 2014; Ford, 2015; Kaplan, 2015). The more such works began to shape my gaze, the more questions I had about how such ideas might be translated to explain what I was grappling with in South Africa (awkwardly called "technologization" in the grant, because the use of trucks instead of carts for recyclables seemed to be poorly captured by conventional uses of the term automation).

I chased the question of automation across the mines and farms of South Africa, through explanations in economics for "jobless growth." The deeper I looked, the less I saw the South African economy in terms of labour and wages or growth and employment. The more I came to see it in terms of machines, disemployment and distribution. In the process, I came to reread the outline of South African history, to see not just the ecological, but the sociomaterial lens that was typically left out of the story.

One of the most central concerns for most of South Africa's colonial and apartheid economy was how to attract labour both from rural areas and other countries. This point is uncontested; it is foundational to the long-standing historical narrative (see Cooper, 1996; Barchiesi, 2011, 2016; Lawhon *et al.*, 2019). There is a profound disjuncture between this history and the present, in which unemployment is widely recognized as the most critical challenge facing South Africa, interwoven with nearly every other social and economic problem. Understandably, the political transition that occurred in a similar timeframe has attracted most scholarly attention, and yet, I believe we would be well-served to think much more about just how profound this shift is. It is not that South Africa produces less than it used to (although there are some changes) but that the economy requires fewer and fewer people per unit of output, a trend that is largely true globally. I also believe we may be well-served to think of this not as a historical anomaly, a problem that can be solved through the right kind of growth, but as a new reality that requires new responses.

Of course, it is too much to detail this argument here: I write this section towards an understanding of unlearning rather than as an argument structured to convince the reader on the specifics of the case! (The interested reader might find use in Pierce *et al.*, 2019 on disemployment and geography, and Lawhon *et al.*, 2019 as well as Barchiesi, 2016 in South Africa.)

At first, recognition of the declining importance of labour was frightening, several steps beyond unease. This tone of fearfulness about the future comes across in some of the global north work on automation too (e.g., Ford, 2015), although much northern scholarship suggests new lines of work will emerge (e.g., Brynjolfsson and McAfee, 2014). Yet at the same time as the monographs noted above were published, a very different line of argument was written from a southern perspective. Hulme *et al.* (2012) provide a compelling overview of cash transfers, framed in a way that accords with the wider arc of this book: "the development revolution from the global south." Ferguson's (2015) *Give a Man a Fish* takes up this line of enquiry in South Africa, asking the reader what happens if we give up on the idea that there is work for everyone. He urges us to let go of the productionist approach to critical politics (and its dual focus on wages and ownership) and instead focus on how we distribute the outputs of the economy (see also Ballard, 2013; Ferguson and Li, 2018). In his discussion of South Africa's social grants, Ferguson pushes us to think more carefully about the naturalization of the relationship between work and income. Specifically, these grants are set up for those conceptualized as unable to work: children, elderly and persons with disabilities. They are not meant for those who ought to work, even if there are no jobs. Why, he asks (and urges the reader to ask), focus on creating jobs rather than distributing the wealth generated by the economy?

Ferguson's work makes a crucial contribution to the literature seeking to push us into new imaginations, yet in my own first encounters with the idea of cash transfers, I felt uneasy with the idea of giving away fish, or cash. It made sense enough in a context of high unemployment, so why did this idea make me, and so many I spoke with as I was grappling with the idea, uneasy? I found an answer in a belief long ago identified by one of the founders of western sociology, Max Weber: the Protestant ethic (Weber, 2013 [1920]; see also Weeks (2011); in what follows I call this the modern work ethic). The realities of South Africa pushed me to confront one of my most deeply rooted beliefs: good people work hard; working hard is part of being a good person. Workers should be rewarded for this work; distribution should be based on work. Despite my Catholic upbringing and unlearning so much of what I inherited through this, my belief in the moral value of work had remained firmly intact.

I went back to South African histories in search of interrogations into the travels of this ethic. I saw the modern work ethic deeply embedded in many independence movements and critical scholarship (including but not only Marxist ones) in and about Africa, with the (male) wage-labourer framed as the pinnacle of liberation. Although many criticized the exploitation of labour, and even the role of the colonial modern work ethics in enabling such exploitation, the merit of the ethic itself has gone largely uncontested (although see Barchiesi, 2007, 2011, 2016; we review this in Lawhon *et al.*, 2019).

I had previously accepted it as both normal, and ethically appropriate, that the national liberation movement valorized work, workers and the creation of work as a key purpose of the economy. And in its contemporary incarnation, it had seemed largely sensible to me in a context of widespread unemployment that well-intentioned small business owners in the recycling industry thought of it as a moral/social responsibility to create jobs. I had found the racial politics here undoubtedly fraught: most of the time white business owners created low-skill, low-pay jobs for black workers (and at times white family members too, and while I did not ask, I have little doubt about the wage differences here), but the process of generating employment seemed largely like a good thing[1]. Digging into the roots of the work ethic left me with two key questions: Was creating a need for labour really an appropriate avenue through which to develop social responsibility? More substantively, if job creation was financially inefficient, was creating jobs prudent, productive, good public policy?

I had come to my research on waste with two options: "preserve bad jobs and make them better" or "make better jobs." Both options retained the link between work and income or livelihood, regardless of whether the work actually efficiently contributed to the economy. Instead, I believe we as a research team, similar to the dominant practice of scholars and policy makers, limited our imaginations to these options because we placed a *moral* value on the work itself. For me, it was the link between work and virtue that had to be unlearned in order to rework the binary between two flawed choices. Understanding the prevalence of the association between work and moral virtue helped me to see why many scholars, activists and politicians continue to limit the solutions to unemployment to creating work.

And yet, while it has not been specifically a focus for substantive research, there is a sense that the modern work ethic has been less pervasive in the global south and that this is already enabling more creative imaginaries. Hulme *et al.* (2012) for example argue that the difference between the north and south in terms of the stigmatization of poverty has enabled cash transfers to be much more prevalent and popular in southern countries; cash transfer programs are often undertaken *despite* the advice of northern development "experts." In a context in which the association between work and virtue carries so much weight, it is politically challenging to raise the possibility that work itself is valorized less in the south; it is reasonable to imagine why few researchers might be bold enough to take on this question directly (i.e., because such research has the potential to reinforce ideas of global south "laziness" rather than displace the existing modern work ethic). And yet, if Hulme *et al.*'s (2012) assertion is true, the rejection of the association between work and virtue, combined with an acceptance that more labour not is economically necessary, has enabled southern actors to imagine and enact a very different set of possibilities for a more just and efficient distribution of wealth.

This "development revolution from the south" may well be catching on in unexpected places. In the midst of my reading and writing for the "Turning Livelihoods to Rubbish" project, the United States elected Donald Trump as its President. Trump did unexpectedly well in areas where manufacturing jobs had recently been lost. While Trump has promised to "bring manufacturing jobs back," presumably from overseas, the mainstream US American media is increasingly highlighting that automation is equally or more responsible for these lost jobs. As in South Africa, manufacturing output in the United States has continued to grow despite shedding jobs (Yang, 2018). One of the presidential candidates in the Democratic Party, Andrew Yang, has centred his political platform around a concern with disemployment from automation, and is advocating universal basic income as a response. As of November 2019, he is one of about ten candidates and polling sixth within the party, surprising many including myself. In short, in this moment of unprecedented skepticism regarding the potential for dignified livelihoods from waged employment and disillusionment with the modernist development narrative, there is a growing conversation in and outside the academy, scattered across spaces and disciplines, attempting to think through what might happen if we accept that the global economy simply does not need the labour of most of us (Gorz, 1985; Weeks, 2011).

As in my work in urban infrastructure, the analytical process I am arguing for here is to separate the goals from the modes of achieving them: work is not the good we seek but a means to an end.

Let me pause in my enthusiasm to note the work that these paragraphs are not intended to do: I do not expect them to convince the average reader to break a belief in the modern work ethic and to denaturalize the relationship between work, virtue and income. (Although I confess to having added more detail than necessary in the hope that these paragraphs might prompt the reader to read more along these lines!) Central to the argument throughout this book is that a postcolonial approach to learning from southern cities calls for much more than a deeper engagement with global south empirics. It calls us to examine the data we collect and how we analyze it. I have argued that scholarship on the southern urban critique might benefit from a deeper analysis of what precisely we mean by provincializing and unlearning theory. We can all surely agree that particular phenomena exist across the north and south, while equally agreeing that northern cities and southern ones tend to look different (without creating a binary or reifying these differences). Sometimes, causal explanations are similar; sometimes they are not. I have encouraged us to break the easy habit of finding cases that are adequately explained through pre-existing frameworks and instead seek what is difficult to explain.

In this chapter, I have worked with the idea that much of what scholars of southern cities are concerned about falls outside theory as conventionally articulated within the academy (although many scholars have long examined such issues). In line with long-standing arguments from postcolonial theory, but largely absent from or implicit in contemporary debates in urban studies, I have urged us to more explicitly investigate the norms that shape what we study, how we study it and how this shapes how we and others seek to intervene in cities. Doing so might also help us better understand why many southern scholars and scholars of southern cities often feel so passionately about the southern urban critique: this critique is not only concerned with how well we explain cities but with the travels of judgements about what a real city is and what a good city is and ought to be. More careful articulation of the process of urban theory making and its components may well help us navigate ongoing intellectual quandaries. It may also enable us to develop new imaginaries of what a city is, how it works, and for whom it works, as well as imaginaries of the cities we aspire to contribute to creating.

Note

1 For an interesting contrast, see Levermann (2019) in which individual responsibility for unemployment was framed as outlandish in the United Kingdom.

Conclusion

Sometimes, I dream I am still sixteen, wearing my uniform, rushing through the halls of my Catholic school unsure what I am looking for. Sometimes, I dream I am almost forty (my age now) and never left Kansas, the inverse of Dickens' Scrooge in *A Christmas Carol* as he watches his alternate life: that other version of me smiles and laughs but with a sense that something is not quite right, that the world is so much more than what I know it to be.

I began writing this conclusion, along with my application for tenure at the University of Oklahoma, in the summer of 2019. At the same time, the US American media (social and otherwise) was reacting to the celebrations of the US American women's soccer team (and particularly Megan Rapionoe) after their World Cup successes. Many responses, especially from women, have emphasized that we ought to shout our accomplishments and not downplay our contributions. I do feel some sense of success in having navigated the academy and found a space for myself and my arguments. In some ways, I do feel some sense of professional success at this moment of my academic career.

And yet, I do not quite know how to wrap up a book that has unfolded as an intermixing of my work and life over the past decades. This book is far from a celebration. These are not ideas that one shouts with confidence from rooftops. They are a declaration of an imperfect process, a record of a struggle. Rereading them reminds me that I still feel like an imposter in the academy, a provincial white girl from nowhere important with an inadequate pedigree, out of place and without much useful to say.

Yet for all that uncertainty, the dreams that I wake most fitful from are those where this life has not happened. This feels akin to what Ahmed (2010) works through in her book about the promise of happiness. I do not think I would have been content pursuing a "happy" life. I have always had too many questions, and been unable to set aside contradictions. I have enjoyed the uneasy. I have found contentment in the unfolding of the process, even if easier choices would likely have been happier ones. I wish I could have come along a more direct route, with fewer stumbles, and particularly fewer of the many unintended offenses made along the way. But I am satisfied with the less happy, less easy path.

So I begin this ending embracing the uneasy, although not exactly advising it. As an advisor I try to help others work through different choices and their consequences rather than presume what is right for others. Investigations such as those in this text are unnerving. They challenge our explanations and call us to question our own beliefs. I am mindful, as noted in the introduction, that too much unease has the potential to be regressive, to prompt defensiveness rather than unlearning.

I am also mindful of the privilege that has enabled me to write. I carry my white privilege in and outside the academy. English is my home language, and I fit the academy in part because writing has always been my most comfortable mode of communication. And, without minimizing these other concerns, for most of the years discussed here I was largely financially and emotionally independent and could spend as much time I wanted on my work, broadly construed. The investigations I have worked through in this book were time-consuming and risky. It would, for example, surely have taken fewer hours to publish findings paralleling many others about the livelihoods of informal waste workers instead of following the threads outlined in Chapter 9. It was professionally risky as it was often unclear to me just how I might publish from these searchings and who I might offend, let alone how to explain what I was working on while on the job market. (The difficulty of talking about current work is always hard for big and sprawling ideas, but it is particularly true when challenging norms that others might hold, whether from a postcolonial perspective or not. Maybe this is why few have sought to question the modern work ethic: are any prospective employers excited by the idea of dismantling the cultural belief that makes us feel good when we work hard?) It was a privilege to be able to take risks with my time and writing, aware that it might shape my employment prospects but not fundamentally concerned with putting food on the table.

For all the limits and risks, this has been truly exciting work. I hope that the text up to now has conveyed to the reader some sense of the joy that accompanied the journey. The outputs from my work, and particularly the most recent interventions on heterogeneous infrastructure configurations and unlearning the modern work ethic, have been positively received, and I am hopeful that they will continue to shape ongoing conversations. Part of what makes them so exciting for me is that they are not so much answers as opening towards different ways of thinking that just might enable a better way of doing things. I have sought to make clear throughout this book that I think the world, and the academy, could use more of the insights generated through uneasy engagements with unlearning and learning anew.

Equally, while less of my focus, we come to knowledge in many ways and I write with respect for those who approach their scholarship through other processes. My hope is that we can use these different approaches productively in tandem, rather than positioning them at odds with each other. The goal is new knowledge, and the mode outlined here is but one way to get there.

I have written this book in the hope of working to clarify and contribute to ongoing conversations about the position and contributions of cities in the global south to urban studies. Myers (In press) has recently and usefully described this as a moment of collective unsettling and unease. My hope is that this unease can be used in a productive way. While it is clear that, for some, it had caused a reactionary defence of long-standing interpretations of the urban and urban theory, Myers also argues that even those most critical of the southern urban critique acknowledge the importance of understanding southern cities and many of the contributions of the most prominent southern urban scholars. My hope is that, for most of us, the unease can spark an interest in unlearning, and a greater acceptance of and success for those seeking to learn anew in and beyond southern cities.

It is only in writing this conclusion that I realized that the chapters work fairly neatly in pairs; that kind of symmetry is but an aesthetically pleasing surprise. The first two chapters frame long-standing questions and their current iterations regarding how scholars might think about southern cities and their relationship to northern theory and urban studies. Roughly fifty years ago, scholars became particularly interested in the question of southern difference in southern cities and began calling for alternative theorizations (McGee, 1971; Robinson, 2006). This call has been repeated and reworked over time, but I argue that something has changed recently that has moved this conversation much closer to the centre of the field, troubling the core of urban studies. With Yaffa Truelove, I argue that, ontologically, spatial categories such as "the global south" have little meaning. And yet, the south is a socially constructed category that remains important to our analysis of how the field of urban studies is conducted and urban theory is produced. In this context, we seek to demonstrate three propositions in the current literature underpinning the importance of studying southern cities: (1) the south is empirically different, (2) the south has different intellectual and vernacular traditions and (3) postcolonial relations require us to examine the production of knowledge. We find the last of these most compelling, including its ability to transcend the difficulties present in the idea of thinking "from" the south. This proposition underpins the approach I have taken throughout this monograph.

After working through the foundations of the southern urban critique, I examine the process through which answers to the urban question have been created, particularly with regard to the development and deployment of different answers to the urban question to African cities. In Chapter 3, I show that many studies of African cities have taken definitions of "city" developed from northern scholarly ideas and cities and used them to define what is (not) city in Africa. They have not looked at African vernacular ideas, nor used African cities, as a basis from which to develop scholarly ideas about what we mean by "city." As a result, contemporary scholars have constructed awkward phrases such as "rural in the urban" that are typically only applied to residents of global south cities. Chapter 4, with Anesu Makina, builds on this argument through an examination of vernacular ideas of urban vocabulary. Understanding the roots of our terminology, as well as conflation between scholarly and vernacular usages, matters because what we identify as "city" or "not city" shapes what we analyze when constructing urban theory. It also shapes how we interact with urban research participants. It also shapes our imaginations of what a city ought to be. Building on empirical data from photo-elicitation interviews in South Africa, we show that there remains a strong conflation in the vernacular between urban, modern, European, white and what is desirable. We also point to ways in which such conflations have long spilled over into scholarly theorizations and politics. The chapter concludes with a brief enquiry into what this might mean for critical politics, including how scholars, activists, policy-makers and residents imagine and participate in city visioning exercises and what the right to the city means if we deploy a vernacular understanding of "the city."

Chapters 5 and 6 point us towards what an urban studies after the southern critique might entail, outlining three distinct and intellectually incompatible potential pathways (1) southern urbanism as a distinct field, (2) a postcolonial world-of-cities approach which suggests the locatedness and limits of all knowledge and (3) a global urban studies. Given the fraught history of colonial modern ways of thinking and their continued salience, in Chapters 5 and 6, I have argued (with, respectively, Lené Le Roux and Yaffa Truelove) against a straightforward integration of southern cities into urban studies as it is currently constituted. I cannot say strongly enough that I wish that in the decades during which scholars have been concerned with the travels of theory we had overcome the difficulties of integrating the south into urban studies. I wish, at least in teaching, we could adopt a more integrated curriculum that, at least from a pedagogical perspective, taught northern and southern cities in similar ways. My argument throughout this book is based on the premise that neither as a world nor as a field of urban studies have we overcome the need for postcolonial thinking. Whether we ever will is a question I cannot answer, but I hope generations to follow are more ambivalent. In the present, then, we need to draw on postcolonial thinking to continue to insist on critical reflections about how knowledge is identified and presented.

The final chapters address the question of how we might grapple with the use of northern theory in southern cases. It has been said repeatedly in this book and elsewhere, the southern urban critique is neither a rejection of generalization nor of theory derived elsewhere. Yet, I say cautiously lest this be misquoted: it is a critique of *some* inappropriate generalizations and *some* inappropriate travels of theory. In this text I have sought to push forward ideas between these two statements in a manner that is readable and applicable if not formulaic. I am hopeful it might also offer some clarity for others already working in and across northern and southern urban conversations.

Chapter 7 demonstrates just how difficult it can be to understand the precise arguments in ongoing debates. This is emphatically not a critique of the scholars reviewed here, and I hope that Chapter 8's examination of my own work makes clear that I too have struggled to adequately and consistently develop lines of argument. It is also worth repeating that these are difficult conversations, loaded with pressures of publication as well as emotions and colonial modern ideas, let alone performed within the constraints of word limits. I seek to tease out, through a selective examination of a small number of studies about gentrification and dispossession, just what types of words these terms are intended to be as well as what authors argue ought to be rejected and retained. Loosely, I try to unbundle theory, to separate empirical phenomenon from causal explanations and normative judgments. My argument is that conflation of what is going on with why it is happening and what we ought to do makes it difficult for readers and participants in the conversations to parse particular lines of dissent. This is not to say we can ever cleanly separate these components of theory: "the global south," for example, remains an important analytic, but it is impossible to separate this analytic from the long-standing normative judgments that position the south as inferior. Part of my intention in urging the unbundling of theory is to help us to more clearly and consistently reflect on when terms ought to be discarded, retained, delimited or expanded. Broadly, I hope that naming this difficulty in identifying the components of critique and how we might rework them might enable clearer conversations within specific debates. It might also enable us to develop a better understanding of when and what components of ideas from elsewhere might be useful.

Finally, I work through a series of my own enquiries in an attempt to demonstrate two specific lines of argumentation in the southern urban critique. In Chapter 8, I work from examples that suggest that there are cases beyond what we see when we deploy northern theory. I do so using my own work on environmental justice and urban appropriations. I encourage the examination of cases that fall outside of northern explanations. Importantly, my own work was developed through long-term engagements in South African cities and a sense that there was much left un(der)explained. This is, however, not primarily an argument about southern difference (as outlined in Chapter 2). I could not have identified either of these "gaps" simply by reading the scholarly literature, nor do I believe that a long-term stay without undergoing processes of explicit unease and unlearning would have enabled me to seek and see what fell outside established literatures.

Chapter 9 builds on the idea of unbundling theory to draw our collective attention to normative aspects of urban theory, for it is my understanding that embedded in the southern urban critique is a frustration that goes beyond a question of understanding and into the realm of judgment. I work through two areas of my own recent work (urban infrastructure and the nexus of work, virtue and the politics of distribution) as a way to point towards the norms that travel often as unexpected passengers with scholarly explanation. While it has not been central to the wider arc of this book, my own journey from voluntourist to academic professor has included many bouts of doubt about my role in imagining and instigating change, whether in the United States or South Africa, training students here, from here, and from many elsewheres. There are many others who have grappled with such questions and provided guidance along the way, as briefly reviewed in Chapter 9. Chapter 9 is based equally on my reading of (appropriate) scholarly angst over this role and the voices of friends, colleagues and research participants whom I have encountered. I am committed, as a scholar, not to deciding what change is right, but thinking with others about how change might happen, and what it might look like. In that spirit, Chapter 9 urges us to be more attentive to the norms that travel (often implicitly) with our theories and shape what and how we study. I urge our attention not only towards articulating the norms that underpin what happens in southern cities, but also towards thinking about what norms we might hold onto and how we might better achieve them. I do not take it lightly to say, in a postcolonial spirit, I hope we collectively can be more open to thinking with southern urban residents about imagining and instigating positive change.

Throughout this book and my own scholarship, I have sought to clarify and contribute to developing the southern urban critique not because, in the end, I am particularly invested in a deeper understanding of our world. Robinson and Roy (2016: 185) usefully argue, "For us, what is at stake is the renewal and vitality of concepts and methodologies of the urban." I agree that the southern urban critique fundamentally challenges concepts and methodologies in urban studies, but let us not stop here. My sense is that Robinson and Roy too do not want us collectively to actually stop here. What is at stake in this debate about the meaning of southern cities and how we think about them is our willingness to re-imagine and shape what a city – north or south – can, and ought, to be.

References

Abu-Lughod, J. (1975) The legitimacy of comparisons in comparative urban studies: a theoretical position and an application to North African cities. *Urban Affairs Review*, 11(1), 13–35.

Achebe, C. (1960) *No Longer at Ease*. London: Heinemann.

Achebe, C. (1958) *Things Fall Apart*. London: Heinemann.

Ahmed, S. (2012) *On Being Included: Racism and Diversity in Institutional Life*. Durham, NC: Duke University Press.

Ahmed, S. (2010) *The Promise of Happiness*. Durham, NC: Duke University Press.

Alda-Vidal, C., Kooy, M. and Rusca, M. (2018) Mapping operation and maintenance: an everyday urbanism analysis of inequalities within piped water supply in Lilongwe, Malawi. *Urban Geography*, 39(1), 104–21.

Allen, D., Lawhon, M. and Pierce, J. (2019) Placing race: understanding black geographies through relational place-making. *Progress in Human Geography*, 43(6), 1001–19.

Anand, N., Gupta, A. and Appel, H. (Eds.) (2018) *The Promise of Infrastructure*. Durham, NC: Duke University Press.

Anand, R., Kothari, S. and Kumar, N. (2016) *South Africa: Labor Market Dynamics and Inequality*. International Monetary Fund Working Paper. Available at: https://www.imf.org/en/Publications/WP/Issues/2016/12/31/South-Africa-Labor-Market-Dynamics-and-Inequality-44078

Angelo, H. and Wachsmuth, D. (2015) Urbanizing urban political ecology: a critique of methodological cityism. *International Journal of Urban and Regional Research*, 39(1), 16–27.

Aplerin, J.P., (no date) World scaled by number of documents in web of science by authors living there. Available at: http://jalperin.github.io/d3-cartogram

Asante, M.K. (1988) *The Afrocentric Idea*. Philadelphia, PA: Temple University Press.

Asher, K. (2013) Latin American decolonial thought, or making the subaltern speak. *Geography Compass*, 7(12), 832–42.

Aylett, A. (2010a) Participatory planning, justice, and climate change in Durban, South Africa. *Environment and Planning A*, 42(1), 99–115.

Aylett, A. (2010b) Conflict, collaboration and climate change: participatory democracy and urban environmental struggles in Durban, South Africa. *International Journal of Urban and Regional Research*, 34(3), 478–95.

Bakker, K. (2003) Archipelagos and networks: urbanization and water privatization in the South. *Geographical Journal*, 169(4), 328–41.

Ballard, R. (2004) Middle class neighbourhoods or 'African kraals'? The impact of informal settlements and vagrants on post-apartheid white identity. *Urban Forum*, 15(1), 48–73.

Ballard, R. (2013) Geographies of development II: cash transfers and the reinvention of development for the poor. *Progress in Human Geography*, 37(6), 811–21.

Banerjee, A., Galiani, S., Levinsohn, J., McLaren, Z. and Woolard, I. (2008) Why has unemployment risen in the new South Africa? *Economics of Transition*, 16(4), 715–40.

Bank, L.J. (2011) *Home Spaces, Street Styles: Contesting Power and Identity in a South African City*. London and New York: Pluto Press.

Baptista, I. (2019) Electricity services always in the making: informality and the work of infrastructure maintenance and repair in an African city. *Urban Studies*, 56(3), 510–25.

Barchiesi, F. (2007) Wage labor and social citizenship in the making of post-apartheid South Africa. *Journal of Asian and African Studies*, 42(1), 39–72.

Barchiesi, F. (2011) *Precarious Liberation: Workers, the State, and Contested Social Citizenship in Postapartheid South Africa*. Albany, NY: SUNY Press.

Barchiesi, F. (2016) The violence of work: revisiting South Africa's 'labour question' through precarity and anti-blackness. *Journal of Southern African Studies*, 42(5), 875–91.

Barnett, C. (2003) Media transformation and new practices of citizenship: the example of environmental activism in post-apartheid Durban. *Transformation*, 51, 1–21.

Barnett, C. and Scott, D. (2007) Spaces of opposition: activism and deliberation in post-apartheid environmental politics. *Environment and Planning*, 39(1), 2612–32.

Baronov, D. (2013) *The Dialectics of Inquiry Across the Historical Social Sciences*. New York: Routledge.

Bates, E.A., McCann, J.J., Kaye, L.K. and Taylor, J.C. (2017) "Beyond words": a researcher's guide to using photo elicitation in psychology. *Qualitative Research in Psychology*, 14(4), 459–81.

Battersby, J. (2017) Food system transformation in the absence of food system planning: the case of supermarket and shopping mall retail expansion in Cape Town, South Africa. *Built Environment*, 43(3), 417–30.

Beavon, K.S.O. (1982) Black townships in South Africa: terra incognita for urban geographers. *South African Geographical Journal*, 64(1), 3–20.

Bell, D. and Jayne, M. (2009) Small cities? Towards a research agenda. *International Journal of Urban and Regional Research*, 33(3), 683–99.

Bernt, M. (2016) Very particular, or rather universal? Gentrification through the lenses of Ghertner and López-Morales. *City*, 20(4), 637–44.

Bledsoe, A. (2017) Marronage as a past and present geography in the Americas. *Southeastern Geographer*, 57(1), 30–50.

Blomley, N. (2008) The spaces of critical geography. *Progress in Human Geography*, 32(2), 285–93.

Boland, A. (2007) The trickle-down effect: ideology and the development of premium water networks in China's cities. *International Journal of Urban and Regional Research*, 31(1), 21–40.

Braun, L. and Kisting, S. (2006) Asbestos-related disease in South Africa: the social production of an invisible epidemic. *American Journal of Public Health*, 96(8), 1386–96.

Brenner, N. (2000) The urban question: reflections on Henri Lefebvre, urban theory and the politics of scale. *International Journal of Urban and Regional Research*, 24(2), 361–78.

Brenner, N., Madden, D.J. and Wachsmuth, D. (2011) Assemblage urbanism and the challenges of critical urban theory. *City*, 15(2), 225–40

Bridge, G. and Watson, S. (2003) City imaginaries. In: Bridge, G. and Watson, S. (Eds.) *A Companion to the City*. London: Blackwell, pp. 7–17.

Broeck, S. and Junker, C. (Eds.) (2014) *Postcoloniality-Decoloniality-Black Critique: Joints and Fissures*. Frankfurt: Campus Verlag.

Bruce, D.D. (1992) WEB Du Bois and the idea of double consciousness. *American Literature*, 64(2), 299–309.

Brynjolfsson, E. and McAfee, A. (2014) *The Second Machine Age: Work, Progress, and Prosperity in a Time of Brilliant Technologies*. New York, NY: WW Norton & Company.

Bullard, R.D. (1990) *Dumping in Dixie: Race, Class, and Environmental Quality*. London: Routledge.

Bunnell, T. (2013) City networks as alternative geographies of Southeast Asia. *TRaNS: Trans-Regional and -National Studies*, 1(1), 27–43.

Bunnell, T. and Maringanti, A. (2010) Practising urban and regional research beyond metrocentricity. *International Journal of Urban and Regional Research*, 34(2), 415–20.

Caillods, F. (2016) Knowledge divides: social science production on inequalities and social justice. In: UNESCO World Social Science Report. Available at: http://unesdoc.unesco .org/images/0024/002458/245825e.pdf

Cameron, R. (2004) Local government reorganization in South Africa. In: Meligrana, J. (Ed.) *Redrawing Local Government Boundaries: An International Study of Politics, Procedures, and Decisions*. Vancouver, BC: UBC Press, pp. 206–26.

Carruthers, J. (1995) *The Kruger National Park: A Social and Political History*. Pietermaritzburg: University of Natal Press.

Castells, M. and Sheridan, A. (1977) *The Urban Question: A Marxist Approach (Social Structure and Social Change)*. London: Edward Arnold.

Castree, N, Chatterton, P.A., Heynen, N., Larner, W. and Wright, M.W. (2010) *The Point is to Change It: Geographies of Hope and Survival in an Age of Crisis*. Hoboken, NJ: John Wiley & Sons.

Chakrabarty, D. (2000) *Provincializing Europe: Postcolonial Thought and Historical Difference*. Princeton, NJ: Princeton University Press.

Chaskalson, M. (1989) Apartheid with a human face: Punt Janson and the origins of reform in township administration, 1972–1976. *African Studies*, 48(2), 101–29.

Chattopadhyay, S. (2012) *Unlearning the City: Infrastructure in a New Optical Field*. Minneapolis, MN: University of Minnesota Press.

Christaller, W. and Baskin, C.W. (1966) *[1933] Central Places in Southern Germany*. Translated from *Die zentralen Orte in Süddeutschland*. Englewood Cliffs, NJ: Prentice-Hall.

Christopherson, S. (1989) On being outside 'the project'. *Antipode*, 21, 83–89.

Cinar, A. and Bender, T. (2007) *Urban Imaginaries: Locating the Modern City*. Minneapolis, MN: University of Minnesota Press.

Clark, M. (2015) Making the case for Black with a capital B. Again. *Poynter* Available at: https://www.poynter.org/reporting-editing/2015/making-the-case-for-black-with-a-capital-b-again/

Clifford, J. (1989) Notes on travel and theory. *Inscriptions*, 5. Available at: https:// culturalstudies.ucsc.edu/inscriptions/volume-5/james-clifford/

Cock, J. and Fig, D. (2000) From colonial to community based conservation: environmental justice and the national parks of South Africa. *Society in Transition*, 31(1), 22–35.

Collyer, F. Connell, R., Maia J. and Morrell, R. (2019) *Knowledge and Global Power: Making New Sciences in the South*. Melbourne, Australia: Monash University Publishing.

Comaroff, J. and Comaroff, J.L. (1987) The madman and the migrant: work and labor in the historical consciousness of a South African people. *American Ethnologist*, 14(2), 191–209.

Comaroff, J. and Comaroff, J.L. (2012) *Theory from the South: Or, How Euro-America Is Evolving Toward Africa*. London: Routledge.

Connell, R. (2007) *Southern Theory: The Global Dynamics of Knowledge in Social Science*. Sydney, Australia: Allen & Unwin; Cambridge: Polity Press.

Cook, G.P. (1992) Khayelitsha: new settlement forms in the Cape Peninsula. In: Smith, D. (Ed.) *The Apartheid City and Beyond: Urbanisation and Social Change in South Africa*. London: Routledge.

Cooper, F. (1983) *Struggle for the City: Migrant Labor Capital and the State in Urban Africa*. Beverly Hills, CA and London: Sage Publications.

Cooper, F. (1996) *Decolonization and African society: The labour Question in French and British Africa*. Cambridge: Cambridge University Press.

Cornea, N.L., Véron, R. and Zimmer, A. (2017) Everyday governance and urban environments: towards a more interdisciplinary urban political ecology. *Geography Compass*, 11(4), e12310. Available at: https://doi.org/10.1111/gec3.12310

Cresswell, T. (2012) *Geographic Thought: A Critical Introduction*. Chichester: Wiley-Blackwell.

Crosby, A.W. (1986) *Ecological Imperialism: The Biological Expansion of Europe, 900-1900*. Cambridge: Cambridge University Press.

Dados, N. and Connell, R. (2012) The global south. *Context*, 11(1), 12–13.

Danius, S. and Jonsson, S. (1993) An interview with Gayatri Chakravorty Spivak. *Boundary 2*, 20(2), 24–50.

Datta, A. (2015) A 100 smart cities, a 100 utopias. *Dialogues in Human Geography*, 5(1), 49–53.

Dear, M. (2006) The postmodern turn. In: Minca, C. (Ed.) *Postmodern Geography: Theory and Praxis*. Oxford, UK, and Malden, MA: Blackwell.

Derickson, K. (2016) On the politics of recognition in critical urban scholarship. *Urban Geography*, 37(6), 824–29.

Dewar, D. (1995) The urban question in South Africa: the need for a planning paradigm shift. *Third World Planning Review*, 17(4), 407.

Dixon, J. and Ramutsindela, M. (2006) Urban resettlement and environmental justice in Cape Town. *Cities*, 23(2), 129–39.

Doshi, S. and Ranganathan, M. (2017) Contesting the unethical city: land dispossession and corruption narratives in urban India. *Annals of the American Association of Geographers*, 107(1), 183–99.

Dovers, S., Edgecombe, R. and Guest, B. (Eds.) (2003) *South Africa's Environmental History: Cases and Comparisons*. Athens: Ohio University Press.

Drivdal, L. and Lawhon, M. (2014) Plural regulation of shebeens (informal drinking places). *South African Geographical Journal*, 96(1), 97–112.

Du Bois, WEB. (1990 [1903]) *The Souls of Black Folk*. New York: Vintage Books/Library of America.

Easthope, A. (1998) Bhabha, hybridity and identity. *Textual Practice*, 12(2), 341–8.

Edensor, T. and Jayne, M. (Eds.) (2012) *Urban Theory Beyond the West: A World of Cities*. London and New York: Routledge.

Ekhator, E.O. (2014) Improving access to environmental justice under the African Charter on human and peoples' rights: the roles of NGOs in Nigeria. *African Journal of International and Comparative Law*, 22(1), 63–79.

Epprecht, M. (2016) *Welcome to Greater Edendale: Histories of Environment, Health, and Gender in an African City*. Vol. 5. Canada: McGill-Queen's Press.

Ernstson, H., Lawhon, M. and Duminy, J. (2014) Conceptual vectors of African urbanism: 'engaged theory-making' and 'platforms of engagement'. *Regional Studies*, 48(9), 1563–77.

Esson, J., Noxolo, P., Baxter, R., Daley, P. and Byron, M. (2017) The 2017 RGS-IBG chair's theme: decolonising geographical knowledges, or reproducing coloniality? *Area*, 49(3), 384–8.

Fanon, F. (1952) *Black Skin, White Masks*. New York: Grove Press

Ferenčuhová, S. (2016) Accounts from behind the curtain: history and geography in the critical analysis of urban theory. *International Journal of Urban and Regional Research*, 40(1), 113–31.

Ferenčuhová, S. and Gentile, M. (2016) Introduction: post-socialist cities and urban theory. *Eurasian Geography and Economics*, 57(4–5), 483–96.

Ferguson, J. (1999) *Expectations of Modernity: Myths and Meanings of Urban Life on the Zambian Copperbelt (Perspectives on Southern Africa)*. Berkeley, CA: University of California Press.

Ferguson, J. (2015) *Give a Man a Fish: Reflections on the New Politics of Distribution*. Durham, NC: Duke University Press.

Ferguson, J. and Li, T.M. (2018) *Beyond the "Proper Job:" Political-Economic Analysis After the Century of Labouring Man*. Working Paper 51. Cape Town: PLAAS, UWC.

Ford, M. (2015) *Rise of the Robots: Technology and the Threat of a Jobless Future*. New York: Basic Books.

Freund, B. (2007) *The African City: A History*. Cambridge: Cambridge University Press.

Furlong, K. and Kooy, M. (2017) Worlding water supply: thinking beyond the network in Jakarta. *International Journal of Urban and Regional Research*, 41(6), 888–903.

Garcia-Ramon, M.D. (2004) The spaces of critical geography: an introduction. *Geoforum*, 35(5), 519–650.

Ghertner, A. (2014) India's urban revolution: geographies of displacement beyond gentrification. *Environment and Planning A*, 46(7), 1554–71.

Ghertner, A. (2015) Why gentrification theory fails in 'much of the world'. *City*, 19(4), 552–63.

Giddens, A. (1979) *Central Problems in Social Theory: Action, Structure, and Contradiction in Social Analysis*. Berkeley, CA: University of California Press.

Gillespie, T. (2016) Accumulation by urban dispossession: struggles over urban space in Accra, Ghana. *Transactions of the Institute of British Geographers*, 41(1), 66–77.

Gilmartin, M. and Berg, L.D. (2007) Locating postcolonialism. *Area*, 39(1), 120–4.

Gilroy, P. (2019) Long read | never again: refusing race and salvaging the human. *New Frame*. Available at: https://www.newframe.com/long-read-refusing-race-and-salvaging-the-human/

Glaser, B.G. and Strauss, A.L. (1967) *The Discovery of Grounded Theory: Strategies for Qualitative Research*. Chicago, IL: Aldine Publishing Company

Goankar, D.P. (2001) On alternative modernities. In: Goankar, D.P., (Ed.) *Alternative Modernities*. Durham, NC: Duke University Press, pp. 1–18.

Godlewska, A. and Smith, N., (Eds.) (1994) *Geography and Empire*. Oxford: Blackwell.

Goebel, A. (2015) *On Their Own: Women, Urbanization, and the Right to the City in South Africa*. Montreal: McGill-Queen's Press.

Gorz, A. (1985) *Paths to Paradise: On the Liberation from Work*. London: Pluto Press.

Graham, S. and Marvin, S. (2001) *Splintering Urbanism. Networked Infrastructures, Technological Mobilities and the Urban Condition*. New York: Routledge.

Graham, S. and McFarlane, C. (Eds.) (2014) *Infrastructural Lives: Urban Infrastructure in Context*. New York: Routledge.

Grosfoguel, R. (2011) Decolonizing post-colonial studies and paradigms of political-economy: Transmodernity, decolonial thinking, and global coloniality. *Transmodernity: Journal of Peripheral Cultural Production of the Luso-Hispanic World*, 1(1), 1–36.

Grove, R.H. (1996) *Green Imperialism: Colonial Expansion, Tropical Island Edens and the Origins of Environmentalism, 1600-1860*. Cambridge: Cambridge University Press.

Hall, S. (2017) *The Fateful Triangle: Race, Ethnicity, Nation*. Cambridge, MA: Harvard University Press.

Hall, T. (2006) *Urban Geography*, 3rd edition. New York: Routledge.

Hall, T. and Barett, H.L. (2012) *Urban Geography*, 4th edition. New York: Routledge.

Hallowes, D. (1993) *Hidden Faces: Environment, Development, Justice: South Africa and the Global Context*. Scottsville, South Africa: Earthlife Africa.

Hannerz, U. (1980) *Exploring the City: Inquiries Toward an Urban Anthropology*. New York: Columbia University Press.

Haraway, D. (1991) *Simians, Cyborgs and Women: The Reinvention of Nature*. New York: Routledge.

Harper, D. (2002) Talking about pictures: a case for photo elicitation. *Visual studies*, 17(1), 13–26.

Hart, G. (2002) Geography and development: development/s beyond neoliberalism? Power, culture, political economy. *Progress in Human Geography*, 26(6), 812–22.

Hart, G. (2018) Relational comparison revisited: Marxist postcolonial geographies in practice. *Progress in Human Geography*, 42(3), 371–94.

Hart, G.P. and Sitas, A. (2004) Beyond the urban-rural divide: linking land, labour, and livelihoods. *Transformation: Critical Perspectives on Southern Africa*, 56(1), 31–38.

Hartman, Y. and Darab, S. (2012) A call for slow scholarship: a case study on the intensification of academic life and its implications for pedagogy. *Review of Education, Pedagogy, and Cultural Studies*, 34(1–2), 49–60.

Harvey, D. (1978) The urban process under capitalism: a framework for analysis. *International Journal of Urban and Regional Research*, 2(1–4), 101–31.

Holifield, R. (2009) Actor-network theory as a critical approach to environmental justice: a case against synthesis with urban political ecology. *Antipode*, 41(4), 637–58.

Hountondji, P.J. (1996) *African Philosophy: Myth and Reality*. Bloomington, IN: Indiana University Press.

Huchzermeyer, M. (2014) Invoking Lefebvre's 'right to the city' in South Africa today: a response to Walsh. *City*, 18(1), 41–49.

Hulme, D., Hanlon, J. and Barrientos, A. (2012) *Just Give Money to the Poor: The Development Revolution from the Global South*. Sterling, VA: Kumarian Press for University of Manchester

Iles, A. (2004) Mapping environmental justice in technology flows: computer waste impacts in Asia. *Global Environmental Politics*, 4(4), 76–107.

Jacobs, J. (1961) *The Death and Life of Great American Cities*. New York: Random House.

Jacobs, J.M. (1996) *Edge of Empire: Postcolonialism and the City*. New York: Routledge.

Jacobs, N.J. (2003) *Environment, Power, and Injustice: A South African History*. Cambridge: Cambridge University Press.

Jaglin, S. (2008) Differentiating networked services in Cape Town: echoes of splintering urbanism? *Geoforum*, 39(6), 1897–906.

Jaglin, S. (2014) Regulating service delivery in southern cities: rethinking urban heterogeneity. In: Parnell, S. and Oldfield, S. (Eds.) *The Routledge Handbook on Cities of the Global South*. London: Routledge, pp. 343–446.

Jankie, D. (2004) "Tell me who you are": problematizing the construction and positionalities of "insider"/"outsider" of a "native" ethnographer in a postcolonial context. In: Mutua, K. and Swadener, B.B. (Eds.) *Decolonizing Research in Cross-Cultural Contexts: Critical Personal Narratives*. Albany, NY: SUNY Press, pp. 87–106.

Jazeel, T. (2017) Mainstreaming geography's decolonial imperative. *Transactions of the Institute of British Geographers*, 42(3), 334–7.

Jazeel, T. (2019a) *Postcolonialism*. New York: Routledge

Jazeel, T. (2019b) Singularity. A manifesto for incomparable geographies. *Singapore Journal of Tropical Geography*, 40(1), 5–21.

Jazeel, T. and McFarlane, C. (2010) The limits of responsibility: a postcolonial politics of academic knowledge production. *Transactions of the Institute of British Geographers*, 35(1), 109–24.

Jonas, A., McCann, E. and Thomas, M. (2015) *Urban Geography: A Critical Introduction*. West Sussex: Wiley Blackwell.

Jones, P.S. (1999) 'To come together for progress': modernization and nation-building in South Africa's Bantustan periphery – the case of Bophuthatswana. *Journal of Southern African Studies*, 25(4), 579–605.

Jones, P.S. (2000) 'The basic assumptions as regards the nature and requirements of a capital city' identity, modernization and urban form at Mafikeng's margins. *International Journal of Urban and Regional Research*, 24(1), 25–51.

Kamete, A.Y. (2013) On handling urban informality in southern Africa. *Geografiska Annaler: Series B, Human Geography*, 95(1), 17–31.

Kaplan, D., Holloway, S. and Wheeler, J. (2014) *Urban Geography*. 3rd edition. Hoboken, NJ: Wiley.

Kaplan, J. (2015) *Humans Need Not Apply: A Guide to Wealth and Work in the Age of Artificial Intelligence*. New Haven, CT: Yale University Press.

Kapoor, I. (2004) Hyper-self-reflexive development? Spivak on representing the third world 'other', *Third World Quarterly*, 25(4), 627–47.

Karpouzoglou, T. and Zimmer, A. (2016) Ways of knowing the waste waterscape: urban political ecology and the politics of wastewater in Delhi, India. *Habitat International*, 54, 150–60.

Kenna, T. (2017) Teaching and learning global urban geography: an international learning-centred approach. *Journal of Geography in Higher Education*, 41(1), 39–55.

Kepe, T. and Ntsebeza, L. (Eds.) (2011) *Rural Resistance in South Africa: The Mpondo Revolts after Fifty Years*. Leiden: Brill.

King, A. (1985) Colonial cities: global pivots of change. In: Ross, R.J. and Telkamp, G.J. (Eds.) *Colonial Cities: Essays on Urbanism in a Colonial Context*. Dordrecht: Springer, pp. 7–32.

King, A.D. (2006) Postcolonial cities, postcolonial critiques. In: Berking, H., Frank, S., Frers, L., Löw, M., Meier, L., Steets, S. and Stoetzer, S. (Eds.) *Negotiating Urban Conflicts: Interaction, Space and Control*. Bielefeld: Transcript Verlag. pp 15–28.

Knox, P. and McCarthy, L. (2012) *Urbanisation: An Introduction to Urban Geography*, 3rd edition. New York: Pearson Education.

Krause, M. (2013) The ruralization of the world. *Public Culture*, 25(2), 233–48.

Kuhn, T. (1962) *The Structure of Scientific Revolutions*. Chicago, IL: University of Chicago Press.

Larner, W. (1995) Theorising 'difference' in Aotearoa/New Zealand. Gender, place and culture. *A Journal of Feminist Geography*, 2(2), 177–90.

Lawhon, M. (2013a) Situated, networked environmentalisms: a case for environmental theory from the South. *Geography Compass*, 7(2), 128–38.

Lawhon, M. (2013b) Dumping ground or country-in-transition? Exploring the relevance of global electronic waste discourses to South Africa. *Environment and Planning C*, 31, 700–15.

Lawhon, M. (2018a) Post-Weinstein academia. *ACME: An International Journal for Critical Geographies*, 17(3), 634–42.

Lawhon, M. (2018b) Welcome to Greater Edendale: histories of environment, health and gender in an African city, by Marc Epprecht. *Review for Southeastern Geographer*, 58(2), 212–14.

Lawhon, M., Ernstson, H. and Silver, J. (2014) Provincializing urban political ecology: towards a situated UPE through African urbanism. *Antipode*, 46(2), 497–516.

Lawhon, M. and Fincham, R. (2006) Environmental issues in the South African media: a case study of the Natal Witness. *Southern African Journal of Environmental Education*, 23, 107–20.

Lawhon, M. and Le Roux, L. (2019) Southern urbanism or a world of cities? Modes of expanding urban geographical textbooks, teaching and research. *Urban Geography*.

Lawhon, M. and Makina, A. (2017) Assessing local discourses on water in a South African newspaper. *Local Environment*, 22(2), 240–55.

Lawhon, M., Millington, N. and Stokes, K. (2019) A labour question for the 21st century: Perpetuating the work ethic in the absence of jobs in South Africa's waste sector. *Journal of Southern African Studies*, 44(6), 1115–31.

Lawhon, M., Nilsson, D., Silver, J., Ernston, H. and Lwasa, S. (2018a) Thinking through heterogeneous infrastructure configurations. *Urban Studies*, 55(4), 720–32.

Lawhon, M., Pierce, J. and Bouwer, R. (2018b) Scale and the construction of environmental imaginaries in local news. *South African Geographical Journal*, 100(1), 1–21.

Lawhon, M., Pierce, J. and Makina, A. (2018c) Provincializing urban appropriation: agonistic transgression as a mode of actually existing appropriation in South African cities. *Singapore Journal of Tropical Geography*, 39(1), 117–31.

Lawhon, M., Silver, J., Ernston, H. and Pierce, J. (2016) Unlearning (un)located ideas in the provincialization of urban theory. *Regional Studies*, 50(9), 1611–22.

Lawhon, M. and Truelove, Y. (2020). Disambiguating the southern urban critique: Propositions, pathways and possibilities for a more global urban studies. *Urban Studies*, 57(1), 3–20.

Lea, J.P. (2006) Terence Gary McGee. In: Simon, D. (Ed.) *Fifty Key Thinkers in Development*. New York: Routledge, pp. 176–81

Lees, L. (2012) The geography of gentrification: thinking through comparative urbanism. *Progress in Human Geography*, 36(2), 155–71.

Lees, L., Shin, H.B. and López-Morales, E. (2016) *Planetary Gentrification*. Cambridge, MA: Polity Press.

Lefebvre, H., Kofman, E. and Lebas, E. (1996) *Writings on Cities*. Oxford, UK, and Malden: Blackwell.

Leitner, H. (2008) New series announcement—urban pulse: emerging issues in cities and urban life. *Urban Geography*, 29(6), 517.

Leitner, H. and Sheppard, E. (2016) Provincializing critical urban theory: extending the ecosystem of possibilities. *International Journal of Urban and Regional Research*, 40(1), 228–35.

Leitner, H. and Sheppard, E. (2018) From Kampungs to Condos? Contested accumulations through displacement in Jakarta. *Environment and Planning A: Economy and Space*, 50(2), 437–56.

Leonard, L. and Pelling, M. (2010) Mobilisation and protest: environmental justice in Durban, South Africa. *Local Environment*, 15(2), 137–51.

Lepawsky, J. and McNabb, C. (2010) Mapping international flows of electronic waste. *The Canadian Geographer*, 54(2), 177–95.

Levermann, A. (2019) Individuals can't solve the climate crisis. Governments need to step up. Available at: https://www.theguardian.com/commentisfree/2019/jul/10/individuals-climate-crisis-government-planet-priority

Little, K. (1975) Prologue: methodological perspective and approach in African urban studies, Legon Family Research Papers, No. 3. Legon: University of Ghana.

Loiseau, B., Sibbald, R., Raman, S.A., Darren, B., Loh, L.C. and Dimaras, H. (2016) Perceptions of the role of short-term volunteerism in international development: views from volunteers, local hosts, and community members. *Journal of Tropical Medicine*, Article ID 2569732, 12 pages.

López -Morales, E. (2015) Gentrification in the global South. *City*, 19(4), 564–73.

Losch, A. (1954) *The Economics of Location*. Translated from the Second Revised Edition by WH Woglow with the Assistance of WF Stolper. New Haven, CT, and London: Yale University Press.

Mabin, A. (1992) Dispossession, exploitation and struggle: an historical overview of South African urbanization. In: Smith, D. (Ed.) *The Apartheid City and Beyond: Urbanization and Social Change in South Africa*. London: Routledge, pp. 12–24.

Mabin, A. (2014) Grounding southern city theory in time and place In: Parnell, S. and Oldfield, S. (Eds.) *The Routledge Handbook on Cities of the Global South*. New York: Routledge, pp 21–36.

Mafeje, A. (1978) Soweto and its aftermath. *Review of African Political Economy*, 5(11), 17–30.

Makhulu, A.M. (2015) *Making Freedom: Apartheid, Squatter Politics, and the Struggle for Home*. Durham, NC: Duke University Press.

Mamdani, M. (1996) *Citizen and Subject: Contemporary Africa and the Legacy of Late Colonialism*. Princeton, NJ: Princeton University Press.

Marden, P. (1992). The deconstructionist tendencies of postmodern geographies: a compelling logic? *Progress in Human Geography*, 16(1), 41–57.

Marx, C. (2011) Long-term city visioning and the redistribution of economic infrastructure. *International Journal of Urban and Regional Research*, 35(5), 1012–25.

Massey, D. (2005) *For Space*. London: Sage.

Mayer, P with contributions by Iona Mayer (1961) *Townsmen or Tribesmen: Conservatism and the Process of Urbanism in a South African City*. Cape Town: Oxford University Press.

Maylam, P. (1990) The rise and decline of urban apartheid in South Africa. *African Affairs*, 89(354), 57–84.

Maylam, P. (1995) Explaining the apartheid city: 20 years of South African urban historiography. *Journal of Southern African Studies*, 21(1), 19–38.

Maylam, P. (2001) *South Africa's Racial Past: The History and Historiography of Racism, Segregation, and Apartheid.* New York: Routledge.

McCarthy, J.J. (1992) Urban geography and socio-political change: retrospect and prospect. In: Rogerson, C.M. and McCarthy, J.J. (Eds.) *Geography in a Changing South Africa: Progress and Prospects.* Cape Town: Oxford University Press.

McDonald, D.A. (Ed.) (2004) *Environmental Justice in South Africa.* Athens: Ohio University Press.

McDonald, D.A. (2008) *World City Syndrome: Neoliberalism and Inequality in Cape Town.* New York: Routledge.

McEwan, C. (2003) Material geographies and postcolonialism. *Singapore Journal of Tropical Geography*, 24(3), 340–55.

McFarlane, C. (2012) Rethinking informality: politics, crisis and the city. *Planning Theory and Practice*, 13, 89–108.

McFarlane, C., Desai, R. and Graham, S. (2014) Informal urban sanitation: everyday life, poverty, and comparison. *Annals of the Association of American Geographers*, 104(5), 989–1011.

McFarlane, C. and Robinson, J. (2012) Introduction—experiments in comparative urbanism. *Urban Geography*, 33(6), 765–73.

McFarlane, C., Silver, J. and Truelove, Y. (2017) Cities within cities: intra-urban comparison of infrastructure in Mumbai, Delhi and Cape Town. *Urban Geography*, 38(9), 1393–417.

McGee, T.G. (1971) *The Urbanization Process in the Third World.* London: G. Bell and Sons, Ltd.

McGee, T.G. (1991) Presidential Address: Eurocentrism in geography–the case of Asian urbanization. *Canadian Geographer/Le Géographe Canadien*, 35(4), 332–44.

McLennan, S. (2014) Medical voluntourism in Honduras: 'helping' the poor? *Progress in Development Studies*, 14(2), 163–79.

Merrifield, A. (2014) *The New Urban Question.* London: Pluto Press.

Mignolo, W.D. and Walsh, C.E. (2018) *On Decoloniality: Concepts, Analytics, Praxis.* Durham, NC: Duke University Press.

Millington, N. and Lawhon, M. (2019) Geographies of waste: conceptual vectors from the global south. *Progress in Human Geography*, 43(6), 1044–63.

Miraftab, F. and Kudva, N. (Eds.) (2014) *Cities of the Global South Reader.* New York: Routledge.

Mitchell, J.C. (1954) Urbanization, detribalization and stabilization in Southern Africa: a problem of definition and measurement. Working paper, Available at: http://unesdoc.unesco.org/images/0015/001561/156173eb.pdf

Mitchell, J.C. (1987) *Cities, Society, and Social Perception: A Central African Perspective.* Oxford: Clarendon Press.

Mitchell, K. (1997) Different diasporas and the hype of hybridity. *Environment and Planning D: Society and Space*, 15(5), 533–53.

Monstadt, J. and Schramm, S. (2017) Toward the networked city? Translating technological ideals and planning models in water and sanitation systems in Dar es Salaam. *International Journal of Urban and Regional Research*, 41(1), 104–25.

Moss, P. and Maddrell, A. (2017) Emergent and divergent spaces in the Women's March: the challenges of intersectionality and inclusion. *Gender, Place and Culture*, 24(5), 613–20.

Mountz, A. (2016). Women on the edge: workplace stress at universities in North America. *The Canadian Geographer/Le Géographe canadien*, 60(2), 205–18.

Murray, C. (1987) Displaced urbanization: South Africa's rural slums. *African Affairs*, 86(344), 311–29

Myers, G. (1992) Textbooks and the sociology of scientific knowledge. *English for Specific Purposes*, 11(1), 3–17.

Myers, G. (2001) Introductory human geography textbook representations of Africa. *The Professional Geographer*, 53(4), 522–32.

Myers, G.A. (1994) Eurocentrism and African urbanization: the case of Zanzibar's other side. *Antipode*, 26(3), 195–215.

Myers, G.A. (2003) *Verandahs of Power: Colonialism and Space in Urban Africa*. Syracuse: Syracuse University Press.

Myers, G.A. (2008) Sustainable development and environmental justice in African cities. *Geography Compass*, 2(3), 695–708.

Myers, G.A. (2011) *African Cities: Alternative Visions of Urban Theory and Practice*. London and New York: Zed Books.

Myers, G.A. (2014) From expected to unexpected comparisons: changing the flows of ideas about cities in a post-colonial urban world. *Singapore Journal of Tropical Geography*, 35(1), 104–18.

Myers, G.A. (2015) A world-class city-region? Envisioning the Nairobi of 2030. *American Behavioral Scientist*, 59(3), 328–46.

Myers, G.A. (In press) *Rethinking Urbanism: Lessons from Postcolonial Studies and the Global South*.

Nagar, R (2014) *Muddying the Waters: Coauthoring Feminisms Across Scholarship and Activism*. Urbana, IL: University of Illinois Press.

Newman, K. and Goetz, E. (2016) Reclaiming neighborhood from the inside out: regionalism, globalization, and critical community development. *Urban Geography*, 37(5), 685–99.

Nilsson, D (2019) Water and sanitation is rock 'n' roll. *Välkommen till KTH* Available at: https://www.kth.se/blogs/water/2019/03/water-and-sanitation-is-rock-roll/

Nilsson, D. and Nyanchaga, E.N. (2008) Pipes and politics: a century of change and continuity in Kenyan urban water supply. *The Journal of Modern African Studies*, 46(1), 133–58.

Noxolo, P. (2009) "My paper, my paper": reflections on the embodied production of postcolonial geographical responsibility in academic writing. *Geoforum*, 40(1), 55–65.

O'Connor, A. (1983) *The African City*. New York: Routledge.

Oluo, I. (2018) *So You Want To Talk About Race*. New York: Seal Press.

Pacione, M. (2009) *Urban Geography: A Global Perspective*, 3rd edition. London: Routledge.

Parnell, S. and Oldfield, S. (Eds.) (2014) *The Routledge Handbook on Cities of the Global South*. New York: Routledge.

Parnell, S. and Pieterse, E. (2010) The 'right to the city': institutional imperatives of a developmental state. *International Journal of Urban and Regional Research*, 34(1), 146–62.

Parnell, S. and Robinson, J. (2006) Development and urban policy: Johannesburg's city development strategy. *Urban Studies*, 43(2), 337–55.

Parnell, S. and Robinson, J. (2012) (re)Theorizing cities from the global south: looking beyond neoliberalism. *Urban Geography*, 33(4), 593–617.

Patel, Z. (2006) Africa: A continent of hope? *Local Environment*, 11(1), 7–15.

Peet, R. (1998) *Modern Geographic Thought*. Oxford: Blackwell Publishers.

Pellow, D.N. (2007) *Resisting Global Toxics: Transnational Movements for Environmental Justice*. Cambridge, MA: MIT Press.

Pierce, J. and Lawhon, M. (2015) Walking as method: toward methodological forthrightness and comparability in urban geographical research. *The Professional Geographer*, 67(4), 655–62.

Pierce, J. and Lawhon, M. (2016) What do you mean when you say "urban"? Divergence between everyday language and northern analytical vocabularies in South African Cities. *DIE ERDE–Journal of the Geographical Society of Berlin*, 147(4), 284–9.

Pierce, J., Lawhon, M. and McCreary, T. (2019) From precarious work to obsolete labour? Implications of technological disemployment for geographical scholarship. *Geografiska Annaler Series B*, 101(2), 84–101.

Pieterse, E. (2002) From divided to integrated city? *Urban Forum*, 13(1), 3–37.

Pieterse, E. (2011) Grasping the unknowable: coming to grips with African urbanisms. *Social Dynamics*, 37(1), 5–23.

Pieterse, E (2012) High wire acts: knowledge imperatives of southern urbanisms. *The Johannesburg Salon*, 5, 37–50.

Pieterse, E. (2013) *Epistemic Practices of Southern Urbanism*. 2013 International Journal of Urban and Regional Research (IJURR) Lecture at the Annual Conference of the Association of American Geographers, Los Angeles, April. Available at: http://www.ijurr.org/lecture/2013-ijurr-aag-lecture-epistemic-practices-southernurbanism-edgar-pieterse/

Pihljak, L.H., Rusca, M., Alda-Vidal, C. and Schwartz, K. (2019) Everyday practices in the production of uneven water pricing regimes in Lilongwe, Malawi. *Environment and Planning C: Politics and Space*, Available at: https://doi.org/10.1177/2399654419856021.

Popke, E.J. and Ballard, R. (2004) Dislocating modernity: identity, space and representations of street trade in Durban, South Africa. *Geoforum*, 35(1), 99–110.

Porter, L. (2004) Unlearning one's privilege: reflections on cross-cultural research with indigenous peoples in South-East Australia. *Planning Theory & Practice*, 5(1), 104–9.

Porter, L. (2016) *Unlearning the Colonial Cultures of Planning*. New York: Routledge.

Potts, D (2010) *Circular Migration in Zimbabwe & Contemporary Sub-Saharan Africa*. Suffolk: James Currey.

Powdermaker, H. (1963) Townsmen or tribesmen: conservatism and the process of urbanization in a South African city, by Philip Mayer. *Review for American Anthropologist*, 65(2), 472–4.

Pycroft, C. (2000) Democracy and delivery: the rationalization of local government in South Africa. *International Review of Administrative Sciences*, 66(1), 143–59.

Radcliffe, S.A. (2017) Decolonising geographical knowledges. *Transactions of the Institute of British Geographers*, 42(3), 329–33.

Radhakrishnan R (2008) Is translation a mode? *European Journal of English Studies*, 12(1), 15–31

Raghuram, P. and Madge, C. (2006) Towards a method for postcolonial development geography? Possibilities and challenges. *Singapore Journal of Tropical Geography*, 27(3), 270–88.

Ramogale, M. (2019) Decolonise the curriculum for global relevance, *Mail and Guardian*. Available at: https://mg.co.za/article/2019-06-26-00-decolonise-the-curriculum-for-global-relevance

Ramutsindela, M. (2007) Resilient geographies: land, boundaries and the consolidation of the former bantustans in post-1994 South Africa. *Geographical Journal*, 173(1), 43–55.

Reintges, C. (1992) Urban (mis) management? A case study of the effects of orderly urbanisation on Duncan Village. In: D. Smith (Ed.) *The Apartheid City and Beyond: Urbanization and Social Change in South Africa.* London: Routledge.

Rigg, J. (2004) *Southeast Asia: The Human Landscape of Modernization and Development,* 2nd edition. London: Routledge.

Robbins, P. (In press) Is less more… or is more less? Scaling the political ecologies of the future. *Political Geography.*

Robinson, J. (2002) Global and world cities: a view from off the map. *International Journal of Urban and Regional Research,* 26(3), 531–54.

Robinson, J. (2004) In the tracks of comparative urbanism: difference, urban modernity and the primitive. *Urban Geography,* 25(8), 709–23.

Robinson, J. (2005) Urban geography: world cities, or a world of cities. *Progress in Human Geography,* 29(6), 757–65.

Robinson, J. (2006) *Ordinary Cities: Between Modernity and Development.* New York: Routledge.

Robinson, J. (2010) Living in dystopia: past, present, and future in contemporary African cities. In: Prakash, G. (Ed.) *Noir Urbanisms: Dystopic Images of the Modern City.* Princeton, NJ: Princeton University Press.

Robinson, J. (2011) Cities in a world of cities: the comparative gesture. *International Journal of Urban and Regional Research,* 35(1), 1–23.

Robinson, J. (2014) New geographies of theorizing the urban: putting comparison to work for global urban studies. In: Parnell, S. and Oldfield, S. (Eds.) *The Routledge Handbook on Cities of the Global South.* New York: Routledge. pp 57–70.

Robinson, J. (2016a) Thinking cities through elsewhere: comparative tactics for a more global urban studies. *Progress in Human Geography,* 40(1), 3–29.

Robinson, J. (2016b) Comparative urbanism: new geographies and cultures of theorising the urban. *International Journal of Urban and Regional Research,* 40(1), 187–99.

Robinson, J. and Roy, A. (2016) Debate on global urbanisms and the nature of urban theory. *International Journal of Urban and Regional Research,* 40(1), 181–6.

Rocheleau, D. and Edmunds, D. (1997) Women, men and trees: gender, power and property in forest and agrarian landscapes. *World Development,* 25(8), 1351–71.

Rogerson, C.M. (1974) Industrialization of the Bantu Homelands. *Geography,* 59(3), 260–4.

Rogerson, C.M. (1990) Sun International: the making of a South African tourismus multinational. *GeoJournal,* 22(3), 345–54.

Roy, A. (2005) Urban informality: toward an epistemology of planning. *Journal of the American Planning Association,* 71(2), 147–58

Roy, A. (2009) The 21st-century metropolis: new geographies of theory. *Regional Studies,* 43(6), 819–30.

Roy, A. (2011) Urbanisms, worlding practices and the theory of planning. *Planning Theory,* 10(1), 6–15.

Roy, A. (2014) Worlding the south: toward a post-colonial urban theory. In: Parnell, S. and Oldfield, S. (Eds.) *The Routledge Handbook on Cities of the Global South.* New York: Routledge. pp. 9–20.

Roy, A (2016) Who's afraid of postcolonial theory? *International Journal of Urban and Regional Research,* 40(1), 200–9.

Roy, A. and Ong, A. (Eds.) (2011) *Worlding Cities: Asian Experiments and the Art of Being Global.* Malden and Oxford: Wiley-Blackwell.

Said, E (1978) *Orientalism*. New York: Pantheon Books

Said, E.W. (1983) *The World, the Text, and the Critic*. Cambridge, MA: Harvard University Press.

Sanders, R. (1992) Eurocentric bias in the study of African urbanization: a provocation to debate. *Antipode*, 24(3), 203–13.

Sayer, A. (1984) Defining the urban. *GeoJournal*, 9(3), 279–85.

Schindler, S. (2014) A New Delhi every day: multiplicities of governance regimes in a transforming metropolis. *Urban Geography*, 35(3), 402–19.

Schindler, S. (2017) Towards a paradigm of Southern urbanism. *City*, 21(1), 47–64.

Schlosberg, D. (2004) Reconceiving environmental justice: global movements and political theories. *Environmental Politics*, 13(3), 517–40.

Scott, A.J. and Storper, M. (2015) The nature of cities: the scope and limits of urban theory. *International Journal of Urban a Regional Research*, 39(1), 1–15.

Scott, D. and Barnett, C. (2009) Something in the air: civic science and contentious environmental politics in post-apartheid South Africa. *Geoforum*, 40(3), 373–82.

Scott, D., Oelofse, C. and Guy, C. (2002) Double trouble: environmental injustice in South Durban. *Agenda*, 52, 50–57.

Seekings, J. (2001) Social ordering and control in the African townships of South Africa: an historical overview of extra-state initiatives from the 1940s to the 1990s. In: Scharf, W. and Nina, D. (Eds.) *The Other Law. Non State Ordering in South Africa*. Cape Town: JUTA Law, pp. 71–97.

Seekings, J. and Keil, R. (2009) The international journal of urban and regional research: an editorial statement. *International Journal of Urban and Regional Research*, 33(2), i–x.

Seekings, J. and Nattrass, N. (2019) South Africa's failure to create manufacturing jobs. *Daily Maverick*. Available at: https://www.dailymaverick.co.za/article/2019-02-19-south-africas-failure-to-create-manufacturing-jobs/

Senghor, L.S. (1966) Negritude: a humanism of the twentieth century. In Grinker, R.R., Lubkemann, S.C. and Steiner, C.B. (Eds.) *Perspectives on Africa: A Reader in Culture, History and Representation*. Malden, MA: John Wiley & Sons, pp. 629–36.

Sharan, A. (2011) From source to sink: 'official' and 'improved' water in Delhi, 1868–1956. *Indian Economic & Social History Review*, 48(3), 425–62.

Sharp, J. (2009) *Geographies of Postcolonialism: Spaces of Power and Representation*. London: Sage.

Shearing, C. (2001) Transforming security: a South African experiment. In: Strang, H. and Braithwaite, J. (Eds.) *Restorative Justice and Civil Society*. Cambridge: Cambridge University Press, pp. 14–34

Shelby, T. (2007) *We Who Are Dark: The Philosophical Foundations of Black Solidarity*. Cambridge, MA: Harvard University Press.

Sheppard, E, Leitner, H. and Maringanti, A. (2013) Provincializing global urbanism: a manifesto *Urban Geography*, 34(7), 893–900.

Smith, N. (1979) Toward a theory of gentrification a back to the city movement by capital, not people. *Journal of the American Planning Association*, 45(4), 538–48.

Sidaway, J.D. (2000) Recontextualising positionality: geographical research and academic fields of power. *Antipode*, 32(3), 260–70.

Sidaway, J.D., Bunnell, T. and Yeoh, B.S.A. (2003) Editor's introduction: geography and postcolonialism. *Singapore Journal of Tropical Geography*, 24(3), 269–72.

Sidaway, J.D. and Hall, T. (2018) Geography textbooks, pedagogy and disciplinary traditions. *Area*, 50(1), 34–42.

Silver, J (2014) Incremental infrastructures: material improvisation and social collaboration across post-colonial Accra. *Urban Geography*, 35(6), 788–804.

Slater, D. (1992) On the borders of social theory: learning from other regions. *Environment and Planning D: Society and Space*, 10(3), 307–27.

Slater, D. (2004) *Geopolitics and the Post-Colonial: Re-thinking North-South Relations.* Malden, Oxford and Victoria: Blackwell Publishing.

Smith, D. (1992). Introduction. In: Smith, D. (Ed.) *The Apartheid City and Beyond.* London: Routledge.

Smith, R.G. (2013) The ordinary city trap. *Environment and Planning A*, 45(10), 2290–304.

Solman, P. (2018) Analysis: how poverty can drive down intelligence. *PBS Newshour.* Available at: https://www.pbs.org/newshour/economy/making-sense/analysis-how-poverty-can-drive-down-intelligence

Sparke, M. (2007) Everywhere but always somewhere: critical geographies of the global south. *The Global South*, 1(1), 117–26.

Sparke, M. (2018) Textbooks as opportunities for interdisciplinarity and planetarity. *Area*, 50(1), 59–62.

Spivak, G (1999) *A Critique of Postcolonial Reason.* Cambridge, MA: Harvard University Press.

Stats SA. (2011) Census 2011 Metadata (Pretoria, South Africa) [WWW document]. Available at: http://www.statssa.gov.za/census/census_2011/census_products/Census_2011_Metadata.pdf (accessed 2 Nov 2017).

Steyn, P. and Wessels, A. (2000) The emergence of new environmentalism in South Africa, 1988-1992. *South African Historical Journal*, 42(1), 210–31.

Tharps, L.L. (2014) The case for Black with a capital B. *The New York Times* Nov. 18.

The New York Times (1983) "Study ties I.Q. scores to Stress" Available at: https://www.nytimes.com/1983/05/31/science/study-ties-iq-scores-to-stress.html

Tilly, C. (1984) *Big Structures, Large Processes, Huge Comparisons.* New York: Russell Sage Foundation.

Tischler, J. (2013) Alliances and partnerships in recycling in Cape Town, South Africa. *Global Studies Working Papers.* Available at: https://publikationen.uni-tuebingen.de/xmlui/bitstream/handle/10900/49883/pdf/GSWP_DA_Jeannine_Tischler.pdf?sequence=1

Tiwale, S., Rusca, M. and Zwarteveen, M. (2018) The power of pipes: mapping urban water inequities through the material properties of networked water infrastructures –the case of Lilongwe, Malawi. *Water Alternatives*, 11(2), 314–35.

Truelove, Y (2016) Incongruent waterworlds: situating the everyday practices and power of water in Delhi. *South Asia Multidisciplinary Academic Journal*, 14.

Truelove, Y (2019) Gray Zones: The Everyday Practices and Governance of Water beyond the Network, *Annals of the American Association of Geographers*, 109(6), 1758–1774

Tuck, E. and Yang, K.W. (2012) Decolonization is not a metaphor. *Decolonization: Indigeneity, Education & Society*, 1(1), 1–40.

Turner, B.L. (2002) Contested identities: human-environment geography and disciplinary implications in a restructuring academy. *Annals of the Association of American Geographers*, 92(1), 52–74.

Turok, I. (1994) Urban planning in the transition from apartheid, part 2: towards reconstruction. *Town Planning Review*, 65(4), 335.

Turok, I. and Watson, V. (2001) Divergent development in South African cities: strategic challenges facing Cape Town. *Urban Forum*, 12(2), 119–38.

Urban Studies Foundation. (undated) URL https://urbanstudiesfoundation.org/funding/international-fellowships/

Visser, G. and Rogerson, C.M. (2014) Reflections on 25 years of urban forum. *Urban Forum*, 25(1), 1–11.

Wa Ngugi, M. (2012) Rethinking the global South. *The Journal of Contemporary Thought* (Summer). Reprinted in globalsouthproject.cornell.edu. Available at: http://www.globalsouthproject.cornell.edu/rethinking-the-global-south.html.

Walker, A. (2004) *In Search of Our Mothers' Gardens: Womanist Prose*. Houghton Mifflin: Harcourt.

Walker, G. (2009) Globalizing environmental justice: the geography and politics of frame contextualization and evolution. *Global Social Policy*, 9(3), 355–82.

Wang, F., Huisman, J., Meskers, C.E., Schluep, M., Stevels, A. and Hagelüken, C. (2012) The best-of-2-worlds philosophy: developing local dismantling and global infrastructure network for sustainable e-waste treatment in emerging economies. *Waste Management*, 32(11), 2134–46.

Watson, V (2009) South: refocusing urban planning on the globe's central urban issues. *Urban Studies*, 46, 2259–75.

Watson, V. (2013) Planning and the 'stubborn realities' of global south-east cities: some emerging ideas. *Planning Theory*, 12(1), 81–100.

Weber, M. (2013 [1920]) *The Protestant Ethic and the Spirit of Capitalism*. New York: Routledge.

Weeks, K. (2011) *The Problem with Work: Feminism, Marxism, Antiwork Politics, and Postwork Imaginaries*. Durham, NC: Duke University Press.

Werbner, P. and Modood, T. (2015) *Debating Cultural Hybridity: Multicultural Identities and the Politics of Anti-racism*. London: Zed Books Ltd.

Williams, G, Meth, P. and Willis, K (2009) *Geographies of Developing Areas: A Global South in a Changing World*. London: Routledge.

Williams, R (1973) *The City and the Country*. New York: Oxford University Press.

Yang, A. (2018). *The War on Normal People: The Truth About America's Disappearing Jobs and Why Universal Basic Income Is Our Future*. New York: Hachette Books.

Yeoh, B.S. (2001) Postcolonial cities. *Progress in Human Geography*, 25(3), 456–68.

Yiftachel, O. (2006) Essay: re-engaging planning theory? Towards 'south-eastern' perspectives. *Planning Theory*, 5(3), 211–22.

Yiftachel, O. (2009) Theoretical notes on gray cities: the coming of urban apartheid? *Planning Theory*, 8(1), 88–100.

Index

Printed in the United States

Printed in the United States
by Baker & Taylor Publisher Services